SPATIAL ECONOMIC ANALYSIS OF TELECOMMUNICATIONS NETWORK EXTERNALITIES

To my parents

Spatial Economic Analysis of Telecommunications Network Externalities

ROBERTA CAPELLO
Dipartimento di Economia e Produzione
Politecnico di Milano
Milan

Avebury

Aldershot · Brookfield USA · Hong Kong · Singapore · Sydney

© R. Capello 1994

All rights reserved. No part of this publication may be reproduced, stored in a retrieval system, or transmitted in any form or by any means, electronic, mechanical, photocopying, recording or otherwise without the prior permission of the publisher.

Published by
Avebury
Ashgate Publishing Ltd
Gower House
Croft Road
Aldershot
Hants. GU11 3HR
England

Ashgate Publishing Company
Old Post Road
Brookfield
Vermont 05036
USA

British Library Cataloguing in Publication Data
Capello, Roberta
 Spatial Economic Analysis of
 Telecommunications Network Externalities
  I. Title
 384.041
ISBN 1 85628 942 7

Library of Congress Cataloging-in-Publication Data
Capello, Roberta, 1962–
 Spatial economic analysis of telecommunications network externalities / Roberta Capello.
  p. cm.
 Includes bibliographical references.
 ISBN 1-85628-942-7 : $67.95 (approx.)
 1. Telecommunication--Economic aspects. 2. Space in economics. 3. Regional economics. I. Title
HE7631.C37 1994
384'.041--dc20
                                    94-19836
                                       CIP

Printed and Bound in Great Britain by
Athenaeum Press Ltd, Newcastle upon Tyne.

Contents

Figures and Tables xi

Acknowledgements xvii

Preface xix

PART A: THEORETICAL REFLECTIONS 1

1. Telecommunications and Economic Transformation 3

1.1. Introduction 3
1.2. The information economy 5
1.3. Structural changes in the telecommunications sector 8
 1.3.1. Pervasiveness 10
 1.3.2. Change in inter-sectoral barriers 11
 1.3.3. National trajectories of technological development 11
1.4. Diffusion mechanisms of telecommunications
 technologies: network externalities 13
1.5. Objectives and outline of the study 14

2. Telecommunications and Network Externality Theories 19

2.1. Introduction 19
2.2. The concept of network externality 19
2.3. A typology of network externalities 25
2.4. Economic implications of the existence of 31
 network externalities for the telecommunications industry

v

      2.4.1. Diffusion processes — 31
      2.4.2. Tariff structure — 37
      2.4.3. Standardisation — 39
      2.4.4. Compatibility and interconnectivity — 41
      2.4.5. Economic and spatial symbiosis — 43
  2.5. Private, club and public goods: the role of network externalities — 45
      2.5.1. Economic definition of different goods: the case of telecommunications networks — 45
      2.5.2. Economic principles — 49
  2.6. The state of the art in the existing literature — 51
  2.7. The contribution of the present work: research issues — 53

3. Regional Aspects of Technological Change — 59

  3.1. Introduction — 59
  3.2. Spatial adoption of new technologies — 63
      3.2.1. Spatial innovation potentials theories — 65
      3.2.2. Spatial adoption potentials theories — 70
  3.3. New technologies and regional performance — 82
      3.3.1. The role of technological changes in regional performance — 82
      3.3.2. Network infrastructure and regional performance — 87
      3.3.3. The role of ICTs in regional disparities — 91
  3.4. Conclusions — 94

4. Industrial and Spatial Performance in the Presence of Network Externalities — 97

  4.1. Introduction — 97
  4.2. Micro-stimuli for telecommunications networks adoption — 98
  4.3. The economic symbiosis concept — 101
      4.3.1. The framework — 101
      4.3.2. Micro-conditions for the exploitation of economic symbiosis — 104
  4.4. The spatial symbiosis concept — 108
      4.4.1. The framework — 108
      4.4.2. Macro-conditions for the exploitation of spatial symbiosis — 110
  4.5. Benefits and costs of network externalities — 112
  4.6. Conclusions — 115

5. A Theoretical Model for Network Externality Analysis 117

5.1. Introduction 117
5.2. Network externalities and microeconomic production theory 118
    5.2.1. The "technology effect" and the "network effect" 118
    5.2.2. A network externality model 120
5.3. A specific network externality model 124
    5.3.1. A simultaneous game with two symmetric agents 124
    5.3.2. A sequential game with two asymmetric agents 127
    5.3.3. Synthesis of results: a game-theoretic approach 130
5.4. A generalisation of the model 133
    5.4.1. A simultaneous game with n firms 133
    5.4.2. A sequential game with n firms: some considerations 141
5.5. The case of telecommunications networks 142
5.6. Limits of the model 144
5.7. Results of the analysis: propositions to be tested 145
5.8. Conclusions 147

PART B: EMPIRICAL INVESTIGATIONS 149

6. An Empirical Framework for Network Externality Measurement: the Case of Italy 151

6.1. Introduction 151
6.2. Telecommunications and regional development: the STAR programme 153
    6.2.1. General features of the programme 153
    6.2.2. The choice of the sample 158
6.3. Methodological framework for the empirical analysis 161
6.4. The database 166
6.5. Conclusions 169

7. Industrial and Regional Variations in Consumption Network Externalities 171

7.1. Introduction 171
7.2. Network externalities as the main reasons for adoption: descriptive results 173

7.2.1. Regional variations	174
7.2.2. Industrial variations	183
7.3. Methodology for an interpretative analysis: the discrete choice modelling approach	188
7.4. Estimated logit model for the regional and industrial level of analysis	191
7.5. Conclusions	195

8. Industrial and Regional Variations in Production Network Externalities — 199

8.1. Introduction	199
8.2. Methodology of analysis	201
8.2.1. The conceptual approach	201
8.2.2. The connectivity index	205
8.2.3. The performance index	208
8.3. Relationship between the adoption of telecommunication networks and the performance of firms and regions	208
8.4. Relationship between the use of telecommunication networks and the performance of firms and regions	214
8.5. A taxonomy of firms: winners and losers	221
8.6. Conclusions	228

9. Micro- and Macro-conditions for Production Network Externality Exploitation — 231

9.1. Introduction	231
9.2. The conceptual model	232
9.3. Methodology of analysis: the causal path analysis	240
9.4. The choice of the variables	241
9.5. Results of the estimated causal path analysis model for the North of Italy	243
9.6. Results of the estimated causal path analysis model for the South of Italy	246
9.7. Conclusions	247

10. Conclusions and Policy Implications — 251

10.1. In retrospect	251
10.2. Results achieved	253

10.3. Policy implications   256
10.4. New research areas   260
    10.4.1. An extension of the present study   260
    10.4.2. Other kinds of network externalities   261
    10.4.3. Other kinds of networks   261
    10.4.4. Other aspects related to network externalities   262
    10.4.5. Different ways to overcome this kind of externalities in the market   263

Annex 1: Mathematical specifications of the model   265

Annex 2: Questionnaire for telecommunications business users   273

Annex 3: Definition of outliers   281

Annex 4: Choice of cluster number   283

References   289

Figures and tables

Figure 1.1.	Outline of the book	16
Figure 2.1.	Effects of network externalities on market equilibrium	22
Figure 2.2.	Demand-supply interaction	24
Figure 2.3.	A typology of network externalities	27
Figure 2.4.	Economic implications in the presence of network externalities	32
Figure 2.5.	The long-run average demand curve	36
Figure 2.6.	Equilibrium in the presence of network externalities	38
Figure 3.1.	Linkage between technological change and regional performance	62
Figure 3.2.	Linkage between technological change and new spatial structure	62
Figure 3.3.	Linkage between innovation and adoption	71
Figure 3.4.	'Creative diffusion' along technological trajectories in space	76
Figure 3.5.	The product life cycle	78
Figure 3.6.	Innovation rates along the product life cycle	79

Figure 3.7.	Evolution of entry barriers in an oligopolistic market	83
Figure 3.8.	Evolution of income disparities in the EC, 1960-1988 (in ECU)	86
Figure 3.9.	Linkage between adoption and greater performance	87
Figure 4.1.	Profitability curve	100
Figure 4.2.	Conceptual framework of network externality effects	101
Figure 4.3.	Relationship between network adoption and corporate performance	105
Figure 4.4.	Economic and spatial symbiosis	109
Figure 4.5.	Benefits and costs curves	113
Figure 5.1.	Effects of network externalities on the production function	123
Figure 5.2.	Possible connectivity among firms	134
Figure 6.1.	Map of objective 1 regions	154
Figure 6.2.	Structure of the STAR programme	157
Figure 6.3.	Structure of telecommunications technologies' impact analysis of economic and spatial symbiosis	161
Figure 7.1.	Structure of chapter 7	173
Figure 8.1.	Structure of chapter 8	200
Figure 8.2.	Undirected graph representing the connectivity among firms	202
Figure 8.3.	Increasing relationship between the degree of connectivity and the corporate and regional performance	203

Figure 8.4.	Relation between the row connectivity index and the performance index at national level	210
Figure 8.5.	Relation between the row connectivity index and the performance index for the North of Italy	213
Figure 8.6.	Relation between the row connectivity index and the performance index for the South of Italy	214
Figure 8.7.	Relation between the weighted connectivity index and the performance index at national level	215
Figure 8.8.	Relation between the weighted connectivity index and the performance index for the North of Italy	218
Figure 8.9.	Relation between the weighted connectivity index and the performance index for the South of Italy	219
Figure 8.10.	Description of clusters for the North of Italy	224
Figure 8.11.	Description of clusters for the South of Italy	226
Figure 9.1.	Causal path analysis of an economic and spatial symbiosis framework	234
Figure 9.2.	A general model for network externality exploitation estimates	237
Figure 9.3.	Estimated path analysis model for the North of Italy	244
Figure 9.4.	Estimated path analysis model for the South of Italy	246
Table 1.1.	Dynamics and implications of the telecommunications sector	10
Table 1.2.	Number of digital lines installed by geographical areas - 1988	12
Table 2.1.	Private, club and public goods: the role of network externalities	47
Table 2.2.	Taxonomy of existing literature on network externalities	52

Table 3.1.	Typology of approaches to regional aspects of technological change	63
Table 3.2.	The product life cycle: spatial aspects of the various stages	75
Table 3.3.	Examples of the production function approach to infrastructure modelling	89
Table 4.1.	Typology of advantages of network externalities	102
Table 4.2.	Typology of benefits and costs of network externalities	114
Table 5.1.	Pay-off matrix in terms of number of contacts in a two firms game on the decision to connect or not	131
Table 5.2.	Pay-off matrix in terms of number of contacts in a firms game on the decision to let one firm enter the network	133
Table 6.1.	Total expenditure on STAR projects - 1993	159
Table 6.2.	Sample structure	167
Table 7.1.	Adopted telecommunications networks and services by macro-areas	174
Table 7.2.	Main reasons for adoption by macro-areas	176
Table 7.3.	Main reasons for non-adoption by macro-areas	178
Table 7.4.	Main conditions for future adoption by macro-areas	179
Table 7.5.	Degree of satisfaction of the quality of communications by macro-areas	180
Table 7.6.	Major persistent user problems in the case of dissatisfaction by macro-areas	181
Table 7.7.	Intensity of relationships after telecommunications technology adoption by macro-areas	182

Table 7.8.	Main reasons in the case of unincreased intensity by macro-areas	182
Table 7.9.	Adopted telecommunications networks and services by industry and by service industry	184
Table 7.10.	Main reasons for adoption by industry and by service industry	185
Table 7.11.	Main conditions for future adoption by industry and by service industry	186
Table 7.12.	Intensity of relationships after telecommunications technology adoption by industry and by service industry	187
Table 7.13	Main reasons in the case of unincreased intensity by industry and by service industry	188
Table 7.14.	Definition of variables	189
Table 7.15.	Degree of dependency between the dependent variable and the categorical variables	191
Table 7.16.	Estimated logit model with respect to the willingness to adopt at the regional level of analysis	192
Table 7.17.	Estimated logit model with respect to the willingness to adopt at the industrial level of analysis	194
Table 8.1.	Correlation coefficients between the row connectivity index and the performance index by macro-areas	211
Table 8.2.	Correlation coefficients between the weighted connectivity index and the performance index by macro-areas	217
Table 8.3.	Average population values for each observations in the North of Italy	223

Table 8.4.	Average population values for each observations in the South of Italy	227
Table 9.1.	Organisational changes at intra-corporate, inter-corporate and spatial levels	235

Acknowledgements

The telecommunications industry had already been the subject of different research investigations which I carried out at the Bocconi University and at the Centre for Urban and Regional Development Studies. Finally, this work has been carried out in collaboration with the Regional Economics Department of the Free Univeristy of Amsterdam, where I had the chance to develop my studies further.

This work has greatly benefited from comments and suggestions of many people. In this context, I am particularly indebted to Prof. Peter Nijkamp of the Free University of Amsterdam and Prof. Roberto Camagni of the Politecnico of Milan. The complementary constructive criticism and observations of my two "teachers", reflecting their extremely different personal characters, stimulated my work and my sense of criticism and creativeness. From both of them I have always had a vivid example of how much enthusiasm and happiness one can put in a scientific work; from both of them I have always received, whenever I needed, scientific and psychological support to carry out this long research project.

In this context, special thanks go also to Prof. Piet Rietveld of the Free University of Amsterdam and Dr. Andrew Gillespie of the University of Newcastle, for support and suggestions in important stages of the work.

Research funds from the Italian National Research Council (CNR) (projects nos. 92.01858.PF74-Trasporti and 92.03266.CT11, directed by Prof. Roberto Camagni) are greatly acknowledged. This source provided the financial support for the conceptual work, and the publication of the study. Part of the empirical analysis was financed with EC funds from the project "The European Evaluation of the STAR Programme". Dutch funds from the Economic and Social Institute of the Free University supported the logistics of the author, and are therefore acknowledged with great appreciation.

I have a great debt of gratitude to many other people, friends and colleagues. First of all, to my Italian colleagues. Over recent years they willingly covered for me, during my short but frequent visits to Amsterdam, in respect of our common duties towards the students and our research director. The Economics Department of the Politecnico of Milan deserves special thanks. I became a lecturer there in March 1993. Despite my role as a full time lecturer, the Department left me sufficient time to finish this important project.

In addition, I must thank the research staff of the Regional Economics Department of the Free University. They have always been happy to engage in constructive and profound discussions whenever I "appeared" in Amsterdam. Thanks are also due to Mrs. Patricia Ellman kindly revised my English, which suffered in places from the influence of the Italian language.

Particular thanks are extended to my brother, Edoardo, for his mathematical expertise and for working with me during some week-ends. I would also like to thank Dr. Alberto Saccardi of the Bocconi University for statistical assistance and for dealing with my tension during computer analyses, when something was "going wrong".

A big thank you goes also to my family and friends, who had to deal with my stress during these years. Without their moral support, I would have never completed this book.

Roberta Capello

Milan, April 1994

Preface

The telecommunications industry is currently a hot topic for society at large, enjoying an unprecedented period of exponential growth. In the last two decades, this industry has become very popular in academic circles as a lively area of multidisciplinary research activity, where engineers, economists, geographers and sociologists have found a common interest. The scientific community's strong attraction to the field of telecommunications is mainly rooted in the profound changes taking place at both the technological and the institutional level. Only a few years ago the telecommunications industry could simply be defined as that sector producing and managing the transmission of voice via the telephone service. Nowadays this definition is far too limited since the complex developments of the technological revolution have enabled telecommunications networks and services to transmit not only voice or text, but also data and images in real time and with no geographical constraints. For this reason it is generally recognised that telecommunications technologies have enormous potential to reshape the organisation of our society and the economy, by providing the opportunity for a different geographical location of economic activities and a different division of labour and of functions. The main strategic features of this new "Information Economy" are materialised in the ever increasing power of telecommunications technologies to disseminate data-streams, to manage previously fragmented data, even at long distance, and to shrink the physical distance between economic actors. Moreover, it is on this local, regional, international and global web of "information superhighways" that the competitive advantage of firms and the comparative advantage of regions will increasingly depend.

For this reason, the development of telecommunications technologies is extremely important: those national, regional and industrial systems which will not have the benefits of these technologies in the immediate future will lose much of their economic position and possibilities of economic development. In recent times the recognition of the significance of telecommunications technological change in the emerging Information Economy has spawned many studies in Industrial Economics which attempt to interpret the economic regularities that explain the diffusion mechanisms of these technologies. The demand for telecommunications services has been widely analysed within a conventional framework which stressed the importance of price and revenue elasticity. However, the limits of these analyses have become evident because of the technological revolution in the telecommunications sector in the past two decades. In this perspective, it is now more relevant to study the economic mechanisms associated with the diffusion of new telecommunications technologies and in this field a rather new concept has come centre stage: the concept of *network externalities*. This concept is associated with the fact that the mechanism of interdependent preferences provokes a "bandwagon effect", i.e. a fast cumulative process in the diffusion of these technologies. According to a growing body of literature, the most interesting feature of telecommunications networks is the *interrelation among decision-making processes of different users*. This direct interdependence explains the rhythm of adoption, since the user-value of these technologies is highly dependent on the number of already existing subscribers. In other words, in the diffusion mechanisms of these technologies, *consumption externalities* matter and define the diffusion rate of these technologies. This study goes a step further than the recent literature, by claiming that telecommunications networks are not only governed by *consumption network externalities*, but generate *production network externalities*, since their advantages may be measured in terms of the performance of firms and regions.

The present study provides a contribution in the field of network externalities by conceptualising and testing empirically: i) the existence of self-sustained mechanisms in the diffusion processes of these strategic technologies, ii) the advantages firms and regions gain from the adoption and use of these technologies, and iii) under which micro- and macro-conditions firms and regions can exploit network externalities to an increasingly greater extent.

This area undoubtedly represents a very topical and attractive field of research where in spite of all the interest much work still needs to be done and where the present work provides a relatively new contribution.

Part A
THEORETICAL REFLECTIONS

1 Telecommunications and economic transformation

1.1. Introduction

Recent years have witnessed a renewed interest in the role of technological change as an engine for economic growth. For a long period relatively little attention was paid to technological innovation in economic theory, where technological change was often interpreted simply as "manna from heaven", as an exogenously determined variable in a complex set of interrelated mechanisms and factors. Now, however, an opposite perspective dominates recent economic theory on innovation and economic growth. The interest demonstrated in the 1970s, and particularly in the 1980s, can partly be explained by the unstable economic situation of those decades, both locally and worldwide. During these years it became increasingly difficult to accept the interpretation of innovation as an exogenous factor which justified its existence outside an economic analysis framework - as the traditional neoclassical economic growth theory usually assumed - leaving the explanation of economic dynamics to other forces and factors. The strong economic fluctuations of these years required a shift towards more adequate growth models which would more satisfactorily interpret these fluctuations and would provide a more appropriate analytical framework for the relevant economic and technological phenomena of the times[1].

It is not surprising at all that during the 1980s the so-called "Economics of Innovation" came to the fore, providing meaningful concepts and interpretative power to phenomena which were very difficult to understand within standard neoclassical economic growth theory. Innovation then became the primary object of study, assuming a well-defined role within economic theory and was explained as an endogenous element within the economy, emerging from a behavioural economic framework. For example, the firm's capacity and ability to innovate are not equally distributed among firms; instead, these qualities are functions of a particular characteristic of some specific economic actors in specific areas. The uneven distribution of innovation among firms and regional systems justifies their different rates of growth. In this period many theories emerged, which interpreted innovation and technological change in general as the main impetus for economic development. In particular, these theories focused their attention on the interpretation of economic fluctuations, by interpreting innovation as the main explanation for the birth of a new fifth "Kondratiev cycle". The important role attributed to technological change explains the intensive theoretical and conceptual effort devoted in many theories to the interpretation of the economic conditions which generate or stimulate innovation activities. In this perspective, new Schumpeterian theories on "path-dependency" (Dosi, 1982; Dosi et al., 1988) and on "corporate routines" (Nelson and Winter, 1977 and 1982) emerged, explaining the innovation process as a complex mechanism of *interdependent cumulative learning processes* distributed unevenly among firms and regions.

While the mere existence of the Kondratiev cycle is still a matter of scientific dispute, some common agreement does exist on the existence of some "structural adjustment periods". During such periods economic growth based on some specific industries and firms may stagnate, while future economic growth largely depends on the generation of new markets and new dynamic sectors (Freeman et al., 1982; Grossman and Helpman, 1991). In recent years the capacity to move the economy towards the fifth Kondratiev cycle has been attributed to the new "Information and Communication Technologies" (ICTs), which are regarded as the driving forces pushing society into the so-called *"Information Economy"* (see Section 1.2).

The current "upswing" in the interest in innovation in economic growth theory is also sustained by the major technological changes taking place in the telecommunications industry, which are having a drastic effect on the nature of the services provided by this sector (see Section 1.3). These radical technological changes are typical of the dynamic pressures towards the new "Economy", which is characterised by different competitive rules among firms and different comparative advantages among spatial systems. Thus, during the second part of the 1980s a vast literature has dealt with the role of telecommunications in economic growth, in which this sector appears to be the prime cause of the transformation of economic order, rules, laws and power. These technologies are interpreted as the "competitive weapons" of the 1990s; industrial, regional and national economic systems which do not adopt these technologies in the near future risk losing their position in the international markets[2].

The importance of these technologies for economic development has stimulated analyses and studies on the adoption and diffusion mechanisms of these technologies. The relatively new branch of "Industrial Economics" has generated many interesting studies, emphasising the nature of telecommunications technologies and their intrinsic characteristic to be *interrelated technologies*. Many of these studies have focused attention on the diffusion mechanisms based on interrelated consumer preferences. In this context, the concept of *network externalities* has been identified, with the aim of explaining the economic rules governing the diffusion mechanisms of these strategic technologies (see Section 1.4)[3].

Up to now these two fields of economic theories, *viz. telecommunications[4] as the motor for economic growth on the one hand* and *network externality theory on the other*, seem to be completely separated in the literature. The first field tends to be mainly studied in the framework of regional economic theory: relatively more emphasis is put on the innovation aspect and on the consequences of the economic rate of growth of firms and regions, and on the territorial transformation occurring in economic activities when telecommunications technologies are introduced. By stimulating a shrinking of the spatial distance among economic actors, telecommunications technologies may drive the economy towards a completely different spatial structure. In contrast, the second field is much more studied within the context of industrial economic theory, and concerns the analysis of the diffusion mechanisms among firms on the basis of standard economic rules and concepts, such as the "externality" concept.

This study is an attempt to bring these two fields together, by providing an analysis of network externalities in the telecommunications sector and their effects on corporate and regional performance (see Section 1.5). It can essentially be regarded as part of the general theoretical reflection on the role of telecommunications in economic development, by emphasising the importance of telecommunications for future economic growth. However, the advantages derived from these technologies stem not only from the technological changes taking place in the sector, but also from their nature as interrelated technologies. This is because when a new subscriber joins the network, the marginal costs of his entry are lower than the marginal benefits he creates for people (firms) already networked. This difference between marginal costs and benefits (in favour of the benefits) inevitably reflects on industrial performance and - via multiplicative effects - on regional performance. This study develops this concept from a theoretical, conceptual and empirical point of view.

1.2. The information economy

Our society is gradually but convincingly exhibiting the signs of a transition towards an *Information Economy*. By this term the literature refers to an economy where there is a slow but constant increase of all economic activities associated to the production, distribution and consumption of information. The rapid rise of the service sector - not only for domestic but also for international activities - mirrors the fact that the western world is increasingly marked by a wide variety of communication and interaction patterns ranging from a local up to a global scale. This tendency is, moreover, reinforced by the emergence of the information and telecommunications sectors, also denoted as the New Information Technology (NIT) sector or the Information and Communication Technology (ICT) sector (see also Giaoutzi and Nijkamp, 1988).

The pioneering study of Machlup (1962), followed by Porat (1977), was the first one to stress the significance of a "knowledge-based" economy in those years when Bell (1973) was signalling the emergence of a "services-dominated economy in post-industrial society". From these early works, a series of theoretical and empirical analyses have followed, strengthening the development of an economy governed by different rules and dependent upon different strategic resources. Jonscher (1983), for example, sought to explain the emergence of the "Information Economy" through categorising economic activities into two classes, viz. "production tasks", i.e. tasks associated with the manufacturing and delivery of products and services, and "information tasks", i.e. tasks associated with the coordination and manipulation of production tasks. The major source of added value appeared to be shifted from production tasks to information tasks.

All these studies witness the emergence of an information economy, characterised by a growth and intensification of activities (both investment and employment) associated with the collection, manipulation, storage and transfer of information. Within the economy then, *knowledge-based* and *information-based* activities were becoming important strategic resources upon which the competitiveness of firms and comparative advantages for regions would increasingly depend (Bar et al., 1989; Capello et al., 1990; Gillespie et al., 1989; Gillespie and Williams, 1988). Thus, a modern economy tends to go through a period of transformation, marked by the move from "*capital-intensive*" production systems to "*information-intensive*" production systems (Willinger and Zuscovitch, 1988), where information and knowledge are inextricably linked strategic resources for economic development.

The emergence of the "Information Economy" is seen to be highly dependent upon the widespread diffusion and adoption of new ICTs, originating from the interaction of computing and telecommunications, which give rise to new ways of storing, manipulating, organising and transmitting information.

It should be recognised that this qualitative shift towards a new spatial and organisational configuration of activities is not unique in the history of the western world. Also in the past, various types of drastic restructuring or transformation phenomena have occurred. For instance, Andersson and Stroemqvist (1988) distinguished four major transformations (so-called "logistical revolutions") in the past millennium, each characterised by the emergence, acceptance and adoption of fundamentally new types of infrastructure. The following transitional phases are distinguished:

- the *Hanseatic period* (from the thirteenth century onward) based on an integration of sea and land transport, by which Northern Italy was linked to the European Hanseatic League;
- the *Golden Age* (from the beginning of the sixteenth century) based on the improvement in ship building and navigation techniques, by which Europe was connected with other continents;
- the *Industrial Revolution* (from the beginning of the nineteenth century onward) based on new transport and industrial systems technology, by which new world markets could be created;
- the *Information Revolution* (from the 1970s onwards) based on sophisticated interaction and communications channels, by which knowledge and information transfer was possible on a worldwide scale.

Nowadays, Europe can be regarded as a partly integrated and partly competitive network system of countries, regions and metropolitan areas.

It is evident that a delay in the entry to this network will imply a loss of many opportunities and hence cause significant costs, as is witnessed in a recent study on "missing networks" in Europe (see Maggi and Nijkamp, 1992; Nijkamp and Vleugel, 1993; Nijkamp et al., 1994). This also means that network infrastructure policy is of critical importance for the future competitive position of Europe as a whole, and also of its constituent regions and cities. In this context, the ICT sector plays a crucial role, as it is the vehicle for both European integration and intra-European competitiveness.

The ICT sector can be regarded as a cluster incorporating basic innovations in the Neo-Schumpeterian sense (see Freeman et al., 1982); it represents a techno-economic quantum leap leading to a revolution in socio-economic and technological conditions in our society, on the basis of the micro-electronics sector (including e.g. transistors, integrated circuits, microprocessors, computers, laser technology). In this perspective, the telecommunications sector becomes strategic for the competitiveness of territorial and industrial systems: ICTs thus represent a challenge to economic systems to achieve a greater economic performance. Nevertheless, they also represent a new threat which territorial systems (national, regional, urban) have to face in the near future. The lack of ICT infrastructure may imply that territorial systems may become isolated from development processes and from integration processes characterising modern economies in Europe, most notably the integration among European countries, including Western and Eastern European economies. From this perspective the telecommunications sector and its future development become critical for the understanding of the future economic positions of each country, and thereby of the competitiveness of national, regional and urban industrial systems in the 1990s.

1.3. Structural changes in the telecommunications sector

The idea of the emergence of a new "Information Economy" is based on the major structural changes affecting the telecommunications industry during the past two decades. These changes have encompassed technology, regulatory frameworks, market structures, and industry output. The result of these profound changes is the transformation of this industry's image from that of a public utility offering a limited and monopolistic service and operating in a stable and unchanging market structure, to an industry at the forefront of technological change, providing a host of different services and products, operating in a highly competitive market system and destined to transform societies and economic structures (OECD, 1988a and 1988b). The significance of telecommunications in a modern economy has little to do with the "traditional" telephony function of interpersonal communications. Its significance lies in its "technological convergence" with computing technologies, which has greatly expanded the economic potentialities of telecommunications. By enabling computers to communicate over space, and by enabling people to communicate through computers, the new ICTs, emerging from this "technological convergence", lie at the heart of a compound set of technological and economic transformations conveyed by the term "Information Economy" (Gillespie et al., 1989; Jonscher, 1983; Porat, 1977).

The growing awareness of the changing role of telecommunications in the economy raises a fundamental question about the causes of these structural changes. Although technological dynamics are generally pinpointed as the major causes for the restructuring of the telecommunications sector, an analysis of this transformation process focusing only on the technological aspects would be misleading in trying to conceptualise the broader societal characteristics of the telecommunications sector. In this context, at least four factors can be regarded as prominent causes of the transformation of the sector:

- *technological dynamics*. Although it is not the exclusive reason for change, the technological revolution is certainly playing an important role in the development trajectories of the ICT sector. A host of product innovations is taking place, from digitalisation of switching and transmission equipment to a broad range of new services which offer high transportation possibilities of data, voice, text and images;

- *institutional dynamics*, changing the market structure from a monopoly to a competitive market, imposing new "game rules", after decades of traditional static oligopoly (in the manufacturing firms market) and monopoly regimes (in the service market);
- *demand dynamics*, stemming from an increased awareness of users about the strategic importance of these infrastructures, and stimulated by customers' attempts to steer suppliers towards specific products and innovations, thus acting as "technological gatekeepers";
- *new economic relationships* characterising the telecommunications "filière": the matrix of economic relationships among manufacturing firms, and between suppliers, the operator and customers. The traditional oligopolistic rules which have historically governed manufacturing firms and their linkages with public operators have been substituted in recent decades by more competitive rules, by low national protective barriers and by greater competitive threats coming from firms belonging to previously separated sectors.

All these changes have strong implications, namely (see also Table 1.1):

- *a spatial perspective*, regarding the role new computer networks can have in fostering regional and urban development and in shaping a new spatial division of labour. Despite general beliefs about their capacity to overcome the "tyranny of space", computer networks can in fact lead to a concentration of economic activities in core areas. While the spatial development of computer networks is mainly a centripetal process, these technologies can also generate economic development more easily in central areas rather than in the periphery;
- *a demand perspective*, regarding the additional capacity many users have in exploiting these new technological potentialities. The importance these technologies represent for achieving competitive advantages for users can, on the other hand, turn into a threat imposed on users to race to adopt these technologies in a shorter time span and to grapple with their application for new purposes;
- *a supply perspective*, regarding both the new marketing and competitive strategies suppliers have to put in place in view of the rapid technological, institutional and market dynamics the sector is facing.

Seen from these perspectives, the transition process from telecommunication to computer networks, and thus to the "Information Economy", is far from being a simple development trajectory and is fraught with many difficulties. However, it is undoubtedly true that a radical technological process is governing the telecommunications sector and on the basis of these changes the idea of a metamorphosis of our society into an "Information Economy" can be foreseen.

Table 1.1.
Dynamics and implications of the telecommunications sector

	Spatial implications	Implications for users	Supply implications
Technological dynamics	ICTs can be used to overcome some of the previous problems of peripherality or to strengthen core regions' power	ICTs become the strategic instruments to achieve greater competitive advantage through their innovative use	Increased competition from previously separated sectors
Institutional dynamics	Differences in the geography of service supply	Private networks are more easily developed	Structural changes in each branch
Demand dynamics	Biased existence of ICTs in economically stronger areas and regions	Development of internal learning processes	Interaction between demand and supply: cross-learning processes

From the analysis of the dynamics of this process some crucial and basic features of the technological change in the telecommunications sector can be observed, which can be summarised as a) pervasiveness; b) change in intersectoral barriers; c) national trajectories of technological development (Capello, 1991a). Each of these points is discussed separately below:

1.3.1. Pervasiveness

The technological changes taking place in the telecommunications sector encompass both the manufacturing and the service sectors. As regards the manufacturing sector, all three main branches of the equipment industry (switching, transmission and customer premises equipment) have been widely influenced by radical technological changes, increasing technological potentialities of products and new innovative output. In this field rather radical innovations are embodied in the transition from analogue to digital switching equipment, or from cables and wires to fibre-optics, or, again, from microwave systems to satellites (Monk, 1989; Tolmie, 1987). Even more evident is the host of radical innovations taking place in the service sector, offered on more advanced physical infrastructures and thus capable of transmitting text, voice, image and data.

More interesting than this, however, is the phenomenon of pervasiveness related to the demand side. The diffusion of new information and communication technologies is taking place horizontally in all sectors of the economy, an account of the strategic importance these technologies in the pursuit of a better economic performance. For mature sectors these new technologies represent a means to achieve economic rejuvenation, as is the case for the textile and banking systems. Both sectors exploit product and process innovations through the use of computer networks, despite their highly mature output and standardised production processes (Camagni and Capello, 1990; Camagni and Rabellotti, 1992; Fornengo, 1988; Rullani and Zanfei, 1988). The importance of new technologies is once more evident when we look at high-tech sectors, which greatly benefit from the existence of new technologies as both producers and as users (Antonelli, 1988; Camagni, 1992a).

Pervasiveness is also a feature of the geographical diffusion of these technologies: both core and periphery regions have an interest in developing these technologies. Peripheral regions are supposed to exploit information and capital and thus to need more strategic resources which are not equally distributed over space. By the same token, core areas are interested in infrastructural endowment, even more so if these infrastructures represent the strategic means towards comparative advantages.

1.3.2. Change in intersectoral barriers

Another characteristic of the technological dynamics taking place in the telecommunications sector is its ability to destroy traditional inter-sectoral barriers, and open up competition between previously separated sectors. The software component in new networks and services has risen to a high degree, enabling informatics firms to enter the telecommunications market. For traditional telecommunications firms the threat from these new entrants is exacerbated by two factors (Camagni et al., 1993):

- the lack of technical know-how required to produce new services with high software components;
- the lack of managerial know-how to develop strategic corporate policies in highly competitive markets, originating from years of collusive oligopolistic rules that never stimulated aggressive market policies.

1.3.3. National trajectories of technological development

National responses to these pressures and changes have greatly varied, reflecting in part differences in national market characteristics, and to a large extent differences in the market structure (monopolistic versus competitive markets)[5]. Technological trajectories have thus followed national development patterns. Table 1.2 presents a summary of the development of digital networks in most developed countries, Italy and Spain representing the less advanced of these countries in terms of new technologies development.

Table 1.2.
Number of digital lines installed by geographical areas - 1988

Countries	Total Lines Installed	Total Digital Lines Installed	Potential of Substitution*	Potential of Substitution (in %)
USA	127.2	45.0	82.2	28%
Japan	51.7	9.0	42.7	15%
France	25.8	15.6	10.2	3%
West Germany	23.4	5.5	17.9	6%
Italy	20.0	2.2	17.8	6%
Spain	10.5	0.7	9.8	3%
Sweden	6.1	1.8	4.3	1%
Australia	7.1	1.3	5.8	2%
Argentina	3.2	0.2	3.0	1%
Brasil	8.8	0.7	8.1	3%
Czechoslovakia	2.1	0.1	2.0	3%
China	8.0	0.7	7.3	2%
South-Korea	9.6	1.7	7.9	3%
India	3.5	0.3	3.2	1%
Indonesia	0.8	0.2	0.6	0%
Jugoslavia	3.1	0.2	2.9	1%
Malaya	1.3	0.9	0.4	0%
Mexico	4.3	0.7	3.6	1%
DDR	1.8	0.2	1.6	1%
Taiwan	5.7	0.6	5.1	2%
Hungary	0.8	0.1	0.7	0%
USSR	30.0	0.2	30.1	10%
TOTAL	382.1	89.3	292.8	100%

Values in million of lines
* Total lines installed minus total digital lines installed

Source: Zanfei, 1990

The existing technological asymmetry in advanced networks and services among countries reflects at least different time trajectories of new investments rather than different intentions in investments. Nevertheless, the potential threat of this asymmetry is that it can turn into a discontinuity in the provision of physical networks, and thereby disenfranchise national entities from participation in an "Information Economy" (Charles et al., 1989; Sciberras and Payne, 1986; Williams and Gillespie, 1989). At a lower spatial level this also holds true for regions.

1.4. Diffusion mechanisms of telecommunications technologies: network externalities

In the last few decades, the recognition of the importance of telecommunications technological change in the emerging "Information Economy" has stimulated many studies in Industrial Economics which interpret the economic regularities that explain the diffusion mechanisms of these technologies. The demand for telecommunications services has been widely analysed within a conventional framework which stressed the importance of price and revenue elasticity. The limits of this analysis have become evident because of the technological revolution in the telecommunications sector in the past two decades. The development of new services and more technologically advanced networks requires the examination of some essential elements, which were previously unexplored, namely:

- *dynamic externalities*. The use of new telecommunications products is intrinsically associated with the use of telecommunications services. In other words, the demand for telecommunications services is highly dependent on the number of already existing subscribers of telecommunications services;
- *learning processes*. The new telecommunications services necessarily need a learning period to assimilate how they can be exploited, managed, used and applied to production processes. Thus, these learning processes become a vital element acting on the diffusion mechanisms of telecommunications;
- *technological requirements*. The use and adoption of new technologies necessarily requires a high quality of networks in terms of capacity, reliability and speed of transmission.

As Antonelli (1989) has pointed out, a variety of approaches have been elaborated in order to understand the determinants of technological diffusion processes, focusing on four groups of elements: the macroeconomic situation (the accelerator approach), the epidemic imitation (epidemic approach), the dynamics of supply forces (equilibrium approach) and the role of externalities (externality approach):

a) the *accelerator approach* stresses the role of investments in fixed capital, in which the demand for telecommunications services depends on the levels of investments in new capital goods (telecommunications hardware and software infrastructure) which an economic system is able to sustain[6];

b) the *epidemic approach* (a micro-economic perspective) describes the dynamics of the telecommunications demand in terms of a spread of information from early to late adopters. In this approach collective learning stemming from dynamic economies of scale plays a major role[7];
c) according to the *equilibrium approach*, diffusion is conceptualised as a growth in demand caused by the continual increase in performance of innovated products, which is a traditional supply driven development;
d) the *externality approach* or *new epidemic approach* is based on recent analyses of the determinants of the diffusion of new telecommunications technologies which have highlighted the strategic role of adoption and communications externalities, stressing the importance of each adopter on the user-value of the network[8]. The mechanism of interdependent preferences provokes a "bandwagon effect", i.e. a fast cumulative process in the diffusion of these technologies. According to a growing body of literature, the extremely interesting feature of a telecommunications network is the *interrelation among decision-making processes of different users*. This direct interdependence explains the rhythm of adoption, since the user-value of these technologies is highly dependent on the number of already existing subscribers. In other words, in the diffusion mechanisms of these technologies, so-called *consumption externalities*[9] matter and define the diffusion rate of these technologies.

On the basis of the last observation, many studies have been undertaken, at both a theoretical and an empirical level, which try to conceptualise and prove empirically the validity of this observation. The presence of network externality has some important implications. First, when externalities matter, markets perform very poorly as prices do not reflect all relevant information. Second, when externalities play a role, the coordination of diffusion processes among actors requires augmented signalling devices, such as information exchange based on an ex-ante evaluation of firms, which complement the pricing system (Antonelli, 1992). If network externalities are the reason for entering a network, these implications have to be taken into consideration once infrastructural intervention policies are established in order to support the diffusion of these technologies among industrial and territorial systems. It is in this area of analysis that our work will proceed, as we will explain in more detail in the next section.

1.5. Objectives and outline of the study

The present study may to some extent be positioned in the new epidemic approach, since the main area of analysis is the concept of *network externalities* and its interpretation from a theoretical, conceptual and empirical point of view. In this work a conceptual framework is built on the proposition that network externalities are one of the most important reasons for entering a telecommunications network, thus fully accepting the new principles governing diffusion mechanisms. Since externality mechanisms arise on the demand side and impact on the demand function, they can be labelled "consumption network externalities". However, this study goes a step further than the recent literature, by claiming that telecommunications networks are not only governed by *consumption network externalities*, but also generate *production network externalities*, since the behaviour of a firm linked to a telecommunications network influences the performance of other linked firms and - via multiplicative effects - that of regions.

The positive effects generated by production network externalities on the performance of firms have little to do with the traditional effects on corporate performance generated by innovation processes or economies of scale. Although the effects of innovative processes and economies of scale are similar, the nature of production network externality effects is rather different, because their advantages stem from the difference between the marginal costs and benefits of being networked. This is not true for positive effects generated by innovative processes, or by economies of scale. The former stem from an increase in productivity of production processes, the latter from a decrease in costs resulting from large production dimensions.

While at a conceptual level this distinction may seem simple, at an empirical level the separation of the three effects (i.e. innovative processes effects, economies of scale effects, and network externality effects) in the production function is fraught with difficulties. It is in this particular area that the present study provides a contribution, and tries to separate the network externality effects from the economies of scale effects or the innovative process effects. The way in which this task will be done will be explained in Chapter 8 of this study.

Generally speaking, the study will try to address from a conceptual, empirical and policy point of view the questions:

i) *what is the relative importance of network externalities for firms in their decision to join a network?*
ii) *do firms and regions gain from exploiting network externalities?*
iii) *which are the micro- and macro-conditions stimulating or supporting the exploitation of network externalities?*

The outline of the book is as follows (see also Figure 1.1). First, there is an extensive survey of the literature on the concept of network externalities, and on the role of technological change in regional theories, with the aim of finding as yet unexplored areas of research.

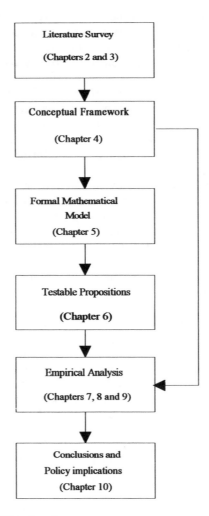

Figure 1.1. Outline of the book

These issues are dealt with in Chapters 2 and 3, respectively. From the literature survey, we are able to develop our own specific conceptual framework ("niche") to explain the role of network externalities on corporate and regional performance (Chapter 4). This conceptual model is then formulated in a formal mathematical model, based on an adjusted neoclassical theoretical framework. The mathematical exercise leads to the formulation of some propositions which then have to be tested in order to judge the empirical validity of our conceptual model (Chapter 5). Once these testable hypotheses have been identified, the empirical analysis is carried out in order to demonstrate the empirical validity of our conceptual framework. In particular, Chapter 6 provides a detailed description of the database used for the Italian case and of the hypotheses to be empirically tested. Chapters 7, 8 and 9 present the empirical analysis, and deal with the basic research questions mentioned earlier in this section. In particular, Chapter 7 deals with the test of consumption network externalities, i.e. it covers empirically the question of whether network externalities are one of the most important reasons for joining a network. Chapter 8 is then devoted to the empirical analysis of production network externalities and deals with the question whether network externalities are exploited by both firms and regions. Finally, Chapter 9 highlights the micro- and macro-conditions which have to be present in order to exploit production network externalities. Chapter 10 presents policy implications arising from our analysis and contains some concluding remarks on future research areas.

Notes

1. For an overview, see Nijkamp and Poot, 1993.
2. The role of ICTs in the emergence of an "Information Economy" has been widely analysed and conceptualised at the Centre for Urban and Regional Development Studies of the University of Newcastle. See, among others, Gillespie et al., 1989; Gillespie and Hepworth, 1986; Gillespie and Williams, 1988; Goddard et al., 1987; Williams, 1987; Williams and Taylor, 1991. The role of information in the performance of firms has been extensively studied by Ouwersloot, 1994.
3. An extensive critical review of the literature on network externality is presented in Chapter 2.
4. In this study the definition "telecommunications technologies" has the same meaning as "Information and Communication Technologies". This latter has been recently used in OECD or EC studies and refers to advanced telecommunication technologies.
5. A vast literature still deals with this debate. To quote just a few examples, see Allen, 1990b; Bradley and Hausman, 1989; Brock, 1981; Camagni and Capello, 1989; Capello, 1991b; Crandall and Flamm, 1989; Curien and Gensollen, 1987; Foreman-Peck and Mueller, 1988; Mueller, 1991; Phillips, 1990; Philips, 1990; Von Weizsaecker and Wieland, 1988.
6. An empirical test of this approach regarding telecommunication infrastructure is presented in Antonelli, Petit and Tahar, 1989.

7. Empirical analyses which follow this approach in the case of the telecommunications sector are described in Antonelli (1990) and Capello (1988).
8. In this study the word "network" refers to a telecommunications network, i.e. a telephone, telex or any advanced digital network transmitting text, data, voice and images. The concept of network externalities is thus related to the telecommunications technologies.
9. In the economic literature, consumption externalities arise when the behaviour of one individual affects (positively or negatively) the consumption of another individual. These externality mechanisms have to be kept separate from the so-called "production externalities"; by this term a situation is meant where the behaviour of an agent affects (positively or negatively) the production of another agent.

2 Telecommunications and network externality theories

2.1. Introduction

The development of the wide technological potentialities and the latest applications in telecommunications has contributed to a revitalised interest in the diffusion and adoption mechanisms of new technologies. Until a few years ago, the demand for telecommunications services was often analysed within a conventional framework which focuses attention on the role of price and revenue elasticity. In recent years however, greater attention has been given to some new mechanisms which are radically influencing the demand for telecommunications services. These mechanisms have been pinpointed by a growing literature as *network externalities* and *learning processes* (Antonelli, 1989).

Since the publication of Rohlf's paper on network externality in 1974, this concept has become the subject of many studies which interpret it as a quintessential issue in the diffusion of new technologies[1]. The aim of the present chapter is to provide a critical review of the literature on network externalities, and to identify as yet unexplored areas of research where our study may provide a new contribution.

2.2. The concept of network externality

The term "*network externality*" stems from the well-known economic concept of externality. In economic theory an externality is said to exist when an external person to a transaction is directly affected (positively or negatively) by the events of the transaction[2]. The concept of "network externality" is related to a simple but fundamental observation that the user-value of a network is highly dependent on the number of already existing subscribers or clients. This means that the choice for a potential user to become a member of the network is dependent on the number of these participants. This basic but crucial statement has strong implications not only for the development trajectories of new networks, but also on some other important elements such as tariff structure, network interconnections, standardisation processes, optimal dimensions of networks and inter-network competition. In other words, the existence of network externality has some far-reaching consequences for the actual operation and policy choices regarding networks. The notion of network externality is thus essentially related to the value of the network, expressed in terms of its subscriber-base.

Recent studies have highlighted the vital role played by network externalities in understanding the environment required for the adoption of innovations and the new capital goods that interlock with them. However, it is still very difficult to give a definition of this concept, as it is subject to many interpretations and easily confused with other phenomena which have little to do with the original concept of external economies. To clarify the concept of network externalities it is useful to keep in mind the two important characteristics attributed to externalities in the economic literature. A first element is *interdependence*, which describes an interaction between the decisions of economic agents. The second is *non-compensation*, so that the one who creates costs (or enjoys benefits) is not obliged to pay for them (Nijkamp, 1977).

In the case of network externalities, the first element, interdependence, is easily identified. The decision of a person to join a network is highly dependent on the number of existing subscribers, i.e. on the number of people who have already made the self-same choice.

More complex is the identification of the second characteristic of externalities, i.e. non-compensation. Useful for this purpose is the distinction between the notion of the cost of purchase and that of the adoption of these technologies. In the case of a telecommunications network, the profitability of these technologies depends only to a limited extent on the prices of the equipment on the market (i.e. the price of fax machines, modems, personal computers to link to networks). Much more relevant are the costs of adoption, such as the learning processes and the organisational changes which firms have to cope with in order to use and exploit these technologies. These costs stem from the behaviour of other firms (which technology they adopt) and on the general level of penetration of the technology in the region (Antonelli, 1991). The higher the number of adopters, the higher the advantage obtained by the technology. This advantage is not incorporated in the cost of purchase, as this cost is not dependent on the number of already existing subscribers. In this sense, the cost of adopting the technology does not reflect all benefits and advantages generated by that technology, and the "non-compensation" element is present. In other words, the actual economic value of telecommunications networks and services is only partially the benefits that individual consumers derive from telecommunications because (Saunder and al., 1983):

a) subscribers may value the service by more than the amount that they are required to pay for it, that is, there might be *consumer surplus* that is not quantified;
b) new telephone subscribers not only incur benefits for themselves, but also increase the benefits of being connected to the system for those who have already joined, that is, there are *subscriber network externalities*;
c) the willingness to pay a given price to make a telephone call reflects only a minimum estimate of the benefits incurred by the caller and does not reflect the benefits received by the recipient of the call or those whom the caller or recipient of the call then contact, that is, there are *call-related externalities.*

As Hayashi (1992) argues, in the telephone network subscribers can benefit either from receiving telephone calls, usually on a free charge basis since the call is charged to the sender, or from enjoying a wide range of call opportunities due to the increasingly large number of subscribers. Both cases are examples of "network externality". The former is often labelled "*call externality*" and excluded from the narrower concept of "network externality". The latter is called "*subscriber externality*" and viewed as the core of the network externality concept (Nambu, 1986; Hayashi, 1992; Wenders, 1987). As explained before, the concept of network externality is related to the advantages which are not paid for once a subscriber joins a network, these advantages being higher the larger the network.

The concept of network externality is then related to the value of the network, which depends on the already existing number of subscribers and differs from the mere cost of purchase of (i.e. access to) the network by that amount of advantage which an individual receives and does not pay for once he joins the network. From this perspective network externalities are the *economic reasons* for the adoption of and entry into the network and are becoming the essential explanation for the diffusion of new interrelated technologies. Firms' decisions to join a new network depend also on the subscriber base of the network and the expectations that potential entrants have of the size of the subscriber base in the near future. Thus, the cost of purchasing the technology itself is not the single element in the decision-making process.

The previous observations can be illustrated by the supply-demand curve in Figure 2.1. In this figure, the individual demand curve on the market, labelled "private marginal benefits", represents the benefits that subscribers receive from joining the network. These benefits are fully paid for by the subscribers via the tariff system of a telecommunications network. If the individual private marginal benefit curve is interpreted in this way, it is too low since not all benefits subscribers can achieve by joining the network are properly represented. The higher demand curve represents the social marginal benefits curve, since it is the result of the sum of the individual marginal benefits curve and the externality curve (Ex) representing the non-paid for advantages subscribers receive from joining the network[3]. In this way network externalities are taken into account, thus shifting the equilibrium point from q^* to $q(1)$, which represents the intersection point between social marginal benefits and marginal costs.

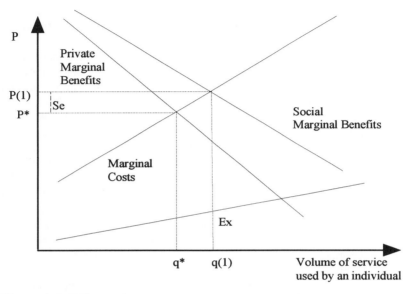

Figure 2.1. Effects of network externalities on market equilibrium

The new price $P(1)$ is higher than P^*, because it is the result of the initial market price of the service (P) and all value added effects of producing the marginal unit of the service, i.e. the aggregate users' value of the new subscriber. Thus $P(1)$ reads as:

$$P(1) = P^* + Se$$

where Se is the value of the new subscriber's effects on the network.

It is clear that optimal output requires an expansion of market output from q^* to $q(1)$ at which point marginal cost is equal to social marginal benefits. The market equilibrium is in this case "disturbed" by the existence of a positive consumption network externality. The diagram in Figure 2.1 is drawn on the basis of an infinite technological capacity of the network, which explains the presence of positive network externalities (i.e. in the chart a positive slope of the externality curve). However, this is not quite realistic. Clearly in the case of fixed networks (i.e. networks with constant technological capacity), *negative consumption network externalities* may arise, as the quality of communications and the rate of failing contacts are affected by congestion (Amiel and Rochet, 1987). Moreover, Figure 2.1 is constructed under the assumption of a competitive market structure. When this is the case, i.e. when competition exists among telecommunications service providers (such as in the case of the United States), the network externality effects are all internalised by consumers. With respect to a monopolistic market structure (such as in the case of most European Countries), the monopolist internalises part of the network externality effects by increasing price and fixing it at a higher value than in the case when network externalities are not present.

The existence of network externality carries an essential message about the economic characteristics of the demand for telecommunications services. As Rohlf (1974) pointed out in his pioneering study, in the presence of network externality the standard micro-economics concept of equilibrium is no longer valid. This assumption stems from the perception of a market demand curve with an increasing slope (measuring the number of subscribers in the network) if network externalities are present in the market. In fact, since the value of the network depends on its interconnectivity and since a large number of subscribers means more possible connections, its value increases with size. In other words, as the quantity of subscribers rises, so does their willingness to pay: that is, a demand price curve which ascends with a positive slope. This situation is presented in Figure 2.2a, which shows the total demand of networking, and not just the individual demand shown in Figure 2.1. Completing Figure 2.2a with a cost curve (supply-price curve) based on economies of scale, we obtain an opposite situation to standard micro-economics (Figure 2.2b). The demand-supply cross, labelled by Allen (1988) the "*Jonscher cross*" from the name of its first inventor, leads to interesting results (see also Roson, 1993):

a) contrary to standard micro-economics, internal market forces drive the system away from equilibrium, and generate a constantly unstable situation. In the first phases of adoption, i.e. when the number of subscribers is low, costs exceed price, thus the supplier faces losses. During maturity phases - with a high number of subscribers - price exceeds cost, thus generating benefits for suppliers. Thus, because of suppliers' incentives to push the system away from the intersection point, dynamic forces lead towards disequilibrium;

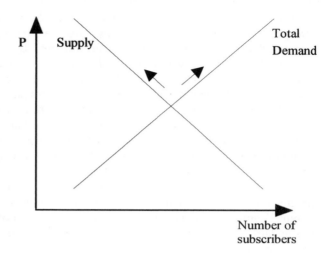

a - Demand-supply interaction in the presence of network externalities

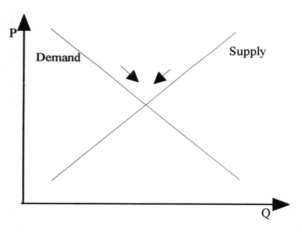

b - Demand-supply interaction in standard micro-economics

Figure 2.2. Demand-supply interaction

Source: Allen, 1988

b) this diagram demonstrates why in the first phases of adoption subsidies are usually necessary because of the expected economic losses. Once the intersection point has been reached and overtaken, economic mechanisms guarantee the diffusion process.

This explanation of network externalities is the most simple and evident framework presented in the literature. However, it hardly begins to explain the complex nature of network externalities, in all their possible manifestations in concrete economic systems.

The simple externality notion presented above is a typical consumption externality. Other types of network externalities are mentioned in the literature, which need to be classified in order to be able to grasp the direct consequences of their existence. In particular, an economic analysis of network externalities and of their consequences requires:

a) *the creation of a typology of network externalities*. The term is often used in the literature related to many different phenomena, and thus generates a range of economic effects. A careful reading of the current literature allows a classification of the concept, based on the distinction between the effects it produces (Sections 2.3 and 2.4 of the present chapter);

b) *a strong relationship with the nature of the good generating the externality*. The private or public nature of the network is in fact an explanatory factor for the existence of network externalities and the economic effects they may produce (Section 2.5 of this chapter).

Once the typology of network externalities and of the nature of network is created, it is easier to understand where the literature has so far concentrated its attention (Section 2.6), and where there is still some scope for new research efforts which would justify the contribution of the present work (Section 2.7).

2.3. A typology of network externalities

The telecommunications industry seems to offer the most appropriate context for studying network externalities and all the economic consequences they provoke. This is because the telecommunications sector is in fact an industry where the concept of externalities, and in particular of network externalities, appears under different guises, which influence both the efficiency of the entire telecommunications system, and, moreover, its dynamics.

As we have seen in the previous section, the concept of network externality is clearly explained when applied to the telephone network. However, a rigorous analysis needs to go far beyond this basic definition, in order to define precisely what is meant by network externalities. Far too broad a definition is given nowadays to this concept, and therefore the need is felt to organise the existing literature dealing with network externalities in a systematic way. Moreover, as we will see below, the typology we create on network externalities shows how this concept is sometimes in reality similar to other more traditional economic concepts. To overcome confusion, a typology of network externalities and their interpretation in the literature is presented in this section.

In the literature on network externality, it is shown how network externalities apply not only to the explanation of demand dynamics, but also to the interpretation of supply mechanisms, driven by interdependent mechanisms. The telecommunications system is in fact characterised by some strategic features, namely:

a) *interdependence of consumer's utility*, since the decision of a person to join the network is dependent on the behaviour of other clients; when specifically dealing with the adoption of a new technology, *interdependence between potential adopters and existing users* exists, since through dynamic learning processes the latter may create for the former a reduction in search costs and market prices for complementary inputs, maintenance and skills which stems from their greater experience in using the technology already adopted;

b) *interdependence between potential users and suppliers*. On the one hand, the know-how and the experience accumulated by suppliers act as a driving force in the adoption process. In fact, the adopting firms are assisted in the search for know-how and complementary inputs (i.e. organisational strategies) because of precise "guidelines" provided by the suppliers. On the other hand, the higher the number of adopters, the broader the know-how of the supply will be. In other words, the relationship between supply and demand generates cross-learning processes via the bridging interaction between demand needs and supply knowledge;

c) *interdependence between producers of complementary technical components and products in the telecommunications "filière"*. The interrelation of sub-markets may provoke externalities, since the profit function of a producer is influenced by the economic transactions of other producers whose behaviour affects the market prices of intermediary inputs;

d) *interdependence between users' productivity*, since the advantages obtained by a firm in terms of its productivity are dependent on the number of already networked firms. In fact, the advantages obtained through the use and exploitation of these technologies are a function of the number of firms already using them.

While the first two of these strategic features (a-b) affect the *utility function of a final individual user*, the last two (c-d) act on the *productivity of firms*, the telecommunications service acting as an input factor in the production function. Moreover, these features are related to both telecommunications manufacturing firms and service providers (the *telecommunications production sphere*) and also to the adopting firms using these technologies as final or intermediate products (the *telecommunications use sphere*). Figure 2.3 presents a typology of network externalities on the basis of the above-mentioned telecommunications market features. In the top-left quadrant of Figure 2.3, network externalities are related to the telecommunications adopters and are the typical *consumption network externalities* acting on the utility function of an individual final user (i.e. the economic features explained in point a). Here the interdependence among utility functions of users of telecommunications networks is at the basis of the traditional network externality concept presented in Section 2.2. Telecommunications demand is more and more explained through interrelated decision-making processes of adopters, a situation which in turn influences the growth rate of telecommunications demand.

Telecommunication use

Consumption network externalities	Pecuniary network externalities
Adoption economies	Technical network externalities

Utility (final use) ←→ Productivity (intermediate use)

| Cross-learning processes | Filière economies |

Telecommunication production

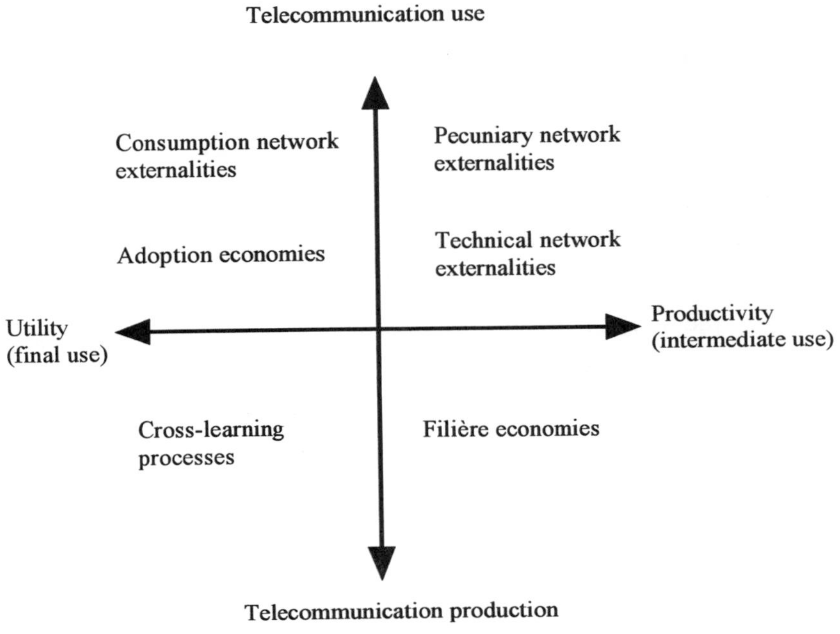

Figure 2.3. A typology of network externalities

A well-known example of consumption network externalities is the so-called hardware /software paradigm (Katz and Shapiro, 1985 and 1986; Stoneman, 1990), regarding the strong interdependent preferences dominating the choice of a consumer when buying a certain kind of hardware. In the words of Katz and Shapiro (1985, p. 424):

".... an agent purchasing a personal computer will be concerned with the number of other agents purchasing similar hardware, because the amount and variety of software which will be supplied for use with a given computer is likely to be an increasing function of the number of hardware units that have been sold".

Also, concerning this example the benefit that a consumer derives from the use of a good is an increasing function of the number and behaviour of other consumers, in this case the fact that they buy compatible items. On the users' side, another kind of network externalities is present, known in the literature as *adoption economies* (Antonelli, 1992), when dealing with the adoption of new technologies (i.e. the economic features explained in point a) above).

In the diffusion processes of new telecommunications technologies a crucial role is also played by collective learning processes, as is common within all types of complex technologies. These processes seem to hide a sort of network externality mechanism, because of non-paid for advantages that potential users of the technology gain from the experience of long-established adopters. For potential adopters, non-paid for advantages may emerge from lower search costs of complementary inputs, or from specific know-how on how to use and maintain the technology, stemming from the consolidated experience on the use of these technologies accumulated by previous adopters.

However, these features, recently interpreted as an externality mechanism (Antonelli, 1992; David, 1992), may in reality be explained only through the traditional concept of dynamic learning processes, which are similar in their effects, but different in nature from the traditional concept of network externalities. Learning processes stem in fact from the concept of dynamic economies of scale (Spence, 1981), while network externalities stem from the non-paid for benefits obtained by interdependent mechanisms. The difference between the two concepts may be more easily explained by recalling the traditional features of externalities mechanisms, i.e. *interdependence* and *non-compensation*. In the case of learning processes the interdependence among users *is* present, and explains part of the diffusion mechanism. The second feature, i.e. non-compensation, is less evident and is what distinguishes learning processes and adoption economies from network externalities. In fact, one may easily argue that even if late adopters may gain from lower search costs for specific know-how on the use of these technologies stemming from consolidated experience of previous adopters, it might well be that:

- these advantages are paid for by late adopters in terms of loss of productivity during the period of non-adoption;
- these advantages may actually be the result of a clear strategy of the first adopter, who could foresee in this behaviour a source of profit, thus eliminating the unintended feature of an externality mechanism, i.e. non-compensation; and, moreover,
- while with network externalities the non-compensation is valid for both the late and the previous adopters, in the case of learning processes the advantages are only in favour of the latter.

The same can be said for the case of telecommunications product firms acting on the utility function of telecommunications users through *cross-learning processes* (bottom-left quadrant in Figure 2.3), linked to the interdependence between potential users and suppliers, described in point b) above. Again, users benefit from these learning processes, through dynamic economies of scale, which are different in nature from the concept of network externalities, for the same reasons explained before.

Network externalities in the telecommunications sector do not only affect the final user, by impacting on his utility function. In the telecommunications industry the intermediate user (or supplier) also acts under certain particular conditions (bottom-right quadrant). As far as the telecommunications technology production is concerned, telecommunications networks are built upon an array of interrelated technical components such as terminals, transmission facilities and switching equipment, as well as intermediate outputs in the extremely complex telecommunications "filière". The interdependence (expressed in point c) above) exists both in vertical relationships (intermediate inputs for telecommunications outputs) as well as in horizontal final products markets (advanced terminals whose development stimulates value added services such as minitel and electronic mail). In both horizontal and vertical interrelationships the behaviour of each economic agent on the market (reduction of prices, new market niches) positively affects the profits of the other interconnected producers, generating what can be interpreted as network externalities. However, these kinds of advantages are typical *filière economies*, stemming from vertical integration in a sector. In other words, these advantages may be associated with traditional "economies of scale" generated in a vertically or horizontally strong market relationship (bottom-right quadrant in Figure 2.3). Another extremely appropriate example of these kinds of "filière economies" is presented by the hardware/software industry. Computers (hardware) and programs (software) have to be used together, and the greater the sales of the hardware are, the higher the profits for software producers will be, via the technical interconnectivity of the two markets.

Finally, an interesting situation concerns interdependence among the productivity of different intermediate users (see point d) above). In this case, it is possible to speak of network externalities, this time related to the use of the service as an input factor for other products, thus having an impact on the productivity level of firms (top-right quadrant in Figure 2.3). In this framework, both the concept of *pecuniary (network) externalities* (Scitovsky, 1954) and *technical (network) externalities* (Meade, 1954) may be useful. Pecuniary externalities arise whenever the profits of one producer are affected by the actions of other producers. In other words, pecuniary externalities act on input factors decreasing their costs and thus having positive effects on the output. This category differs from the "technical external economies", defined by Meade (1954) as those advantages obtained by a firm for its output through the non-paid for exploitation of the output and input factors belonging to other firms. The latter category sees external economies as a peculiarity of the production function, i.e. these external economies act on input factors' productivity. Through the increase in the input productivity these external economies positively influence the corporate output.

For telecommunications network users, the use of the network generates an increase in input productivity (or profit advantages), only partially covered by the costs of joining the network. The non-paid for advantages obtained by a subscriber joining a network have positive effects on the economic performance of the new subscriber. This holds true also for the already existing subscribers, who obtain non-paid for advantages on their production functions if an additional member uses the network. *If network externalities represent one of the (economic) reasons for entering the network, a better economic performance of firms is the (economic) effect they produce on the productivity side.*

From the above observations we conclude that in the telecommunications sector the classical concept of network externality is related only to telecommunications (final or intermediate) users (in Figure 2.3 only the upper half). In recent years, the definition given to network externalities has expanded to embrace network externalities in the production sphere (manufacturing firms and service providers), thus broadening the precise meaning to cover yet more traditional economic concepts.

While consumption network externalities in the context of the use of telecommunications (top-left quadrant in Figure 2.3), as well as adoption economies (learning processes) (bottom-left quadrant) and "filière" economies (bottom-right quadrant) have been widely identified and analysed in the literature, as yet no work has been done on the measurement of the effects of network externalities on the productivity side. The advantages of users joining a network are reflected in the performance of these subscribers via the reduction of input factor costs or the increase in their productivity. These kinds of network externalities and the effects they generate have not yet been investigated. This is the area where the present book provides a contribution by constructing a conceptual and methodological framework dealing with the basic strategic question of whether firms and regions can gain from the network externality effect.

2.4. Economic implications of the existence of network externalities for the telecommunications industry

The creation of a typology of network externalities is fairly useful once the analysis is focused on the economic implications of the existence of network externalities.

Externality mechanisms have always fascinated economists because of their capacity to alter market mechanisms. In the presence of externalities allocation implications arise, which require institutional interventions to compensate for market failure. In the particular case of network externalities, their existence is especially interesting in order to establish the consequences they have for the use and production of telecommunications means. A vast literature has dealt with these implications and with the way in which they may be handled. Figure 2.4 presents a typology of implications on the basis of the previous distinction between the effects they produce on utility and productivity in both the use and production sphere.

This section is devoted to the analysis of the different kinds of implications that may arise in the presence of network externalities. We will first deal with the standard subjects in the literature, i.e. the presence of consumption network externalities for final users of telecommunications technologies and learning processes (top-left quadrant of Figure 2.4). We will then present the case of cross-learning processes and filière economies in the sphere of the production of telecommunications technologies (respectively, bottom-left and bottom-right quadrants of Figure 2.4). Finally, we will consider production network externalities in the case of the intermediate use of telecommunications technologies. This is still a very limited area of study where still much scope for research effort exists, and where the present work attempts to fill part of the knowledge gap.

2.4.1. Diffusion processes

The first and immediate interest in network externalities is related to the effects network externalities have on the diffusion of the demand for telecommunications services and for new technologies.

The common feature of new approaches to the study of telecommunication diffusion processes is that they aim to veer away from the traditional analyses based on price and revenue elasticity, and instead stress that:

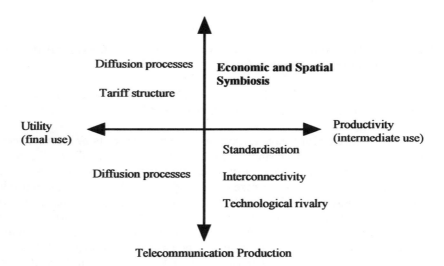

Figure 2.4. Economic implications in the presence of network externalities

a) quality of infrastructure in terms of capacity, speed and reliability is a strategic determinant for telecommunications demand development;
b) current estimates of price elasticities of telecommunications services are likely to play a very reduced role compared with their role in the traditional analyses. In recent approaches, in fact, other determinants, such as the number of existing subscribers or the qualitative level of infrastructure, play a major role in the determinants of demand diffusion.

The acceptance of a network externality approach to the study of telecommunications diffusion processes inevitably raises some analytical issues:

a) the first stems from what has already been said above and concerns the likely relative importance of network externalities in diffusion processes in comparison with other more traditional determinants which influence innovative diffusion processes such as price and income elasticity;
b) a second issue is related to the temporal distribution of network externality effects. In the initial phases, when a small number of subscribers is

connected, the incentive to be connected for an additional subscriber is very low. Thus, in these early stages the network externality effects are not sufficiently strong to generate a self-sustaining process of demand development. In these studies (Allen 1988 and 1990b; Markus 1987 and 1992; Rogers 1986 and 1990) the emphasis has been focused on the concept of "critical mass", i.e. a critical number of subscribers which usually must be built up before the rate of adoption of an interactive technology takes off into rapid growth. In other words, critical mass is the point in time when a sufficient number of individuals has adopted an interactive innovation so that the rate of adoption becomes self-sustained (Allen, 1988; Rogers, 1990).

c) A third issue raised by the existence of network externalities is their effect on the diffusion of telecommunications services over space. The question is whether network externality effects have the same intensity in any locality or if some external conditions and some particular environmental conditions are a necessary prerequisite for network externality effects to play their role in the diffusion processes. In this respect, neither theoretical exercises, nor empirical estimations have been carried out in order to test this possibility.

d) A fourth issue is the evaluation of network externality effects on the diffusion processes of new technologies in different sectors of the economy. It could well be that the ability to appreciate telecommunications services and their economic importance is not distributed equally among the various sectors of the economy.

Another kind of evaluation of network externality effects in different spatial and industrial contexts has up till now been undervalued in the literature. This is because of the difficult task of testing any hypotheses empirically. These are crucial elements in the elaboration of efficient public policies for the development of new technologies. During the start-up phase of new technologies, when network externalities do not play an active role in developing demand (Supply > Demand) (see Figure 2.2a), the effort to overcome this structural inertia of the diffusion process and reach a critical mass will require supply incentives (Allen, 1988). The amount of the incentives, if these are financial incentives, is highly dependent on the critical mass level and this level is in its turn dependent on the interest an individual (or a group of individuals) has in the infrastructure itself. In other words, the amount of the financial incentives required to stimulate networks depends on the intensity with which network externalities play their role in the diffusion process.

In the context of privatisation of provision and use of telecommunications networks some pertinent questions arise, namely:

a) do network externalities play any role in the process of privatisation of telecommunications provision?
b) do network externalities play any role in the diffusion of networks and of services whose use is dedicated to a specific group of people?

The first answer is related to the analyses of the reasons for the transition from monopoly market structures to liberalisation. Many studies have been developed in order to explain these drastic changes, and the main explanations can be summarised as follows:

- *technological reasons*. The most frequent explanation is related to the high-tech revolution taking place in the telecommunications sector. Based on a counter-Schumpeterian theory, this idea is explained by the greater efficiency of a competitive structure to absorb and stimulate technological innovations;
- *political reasons*. Another explanation, complementary to the first one, is the existence in some countries (like Great Britain and Japan) of governments favourable to liberalisation processes. Under these governments, competition is a policy deliberately chosen to enhance efficiency and technological development;
- *endogenous reasons* for network development. A new and very stimulating explanation is provided by Noam (1992), whose approach to the breakdown of public telecommunications systems is related to the intrinsic dynamic nature of telecommunications networks, since their development is governed by consumption network externalities. The very capacity of these internal mechanisms to generate a cumulative rate of adoption is, paradoxically, the cause for the breakdown of a public monopoly. In fact, the continuous stimulus for new entrants to join the network turns into disutility for already existing subscribers, who would accept new entrants up to the point when they have to pay a price for this. In fact, democratising participation leads to democratising the control of cost-sharing in a re-distributory direction, and this re-distributory burden grows as the last participants join the network. After that point, the cohesion of the unitary network breaks apart and new networks are spawned. In Noam's view, in the case of telecommunications the breakdown of the common network is not caused by the failure of the monopolistic system, but rather springs from its very success, i.e. the spread of services across society and the achievement of universality of the services. According to this perception, deregulation should not be seen as a policy of primarily liberalising the entry of suppliers, but rather be interpreted as the liberalisation of exit, by some partners, from a previously existing sharing coalition of users which has become inefficient (Noam, 1987 and 1988).

Noam's approach, although original in its results, is based on some limiting assumptions:

- first, in his model the capacity of the network is fixed, thus leading inevitably to congestion once the number of subscribers increases. On the contrary, even the most traditional network, such as the telephone network, is subject to expansion and technological renewal, continuously increasing its transmission capacity. In this case, the negative effects of network externalities are shifted forward in time so that the explanation for the growing number of private provisions of networks and even private use of networks are not entirely explained through this model;
- secondly, Noam's assumption is that once the common network breaks apart, a number of smaller interconnected networks develop. This can be true for the explanation of the private provision of networks, competitive with the public telephone network. In the case of the private use of a network, however, the interconnection with the public telephone network is not at all guaranteed.

In the case of the use of the network dedicated to a specific group of people, a question is whether network externalities still play the same fundamental role in the diffusion of these technologies as in the presence of public networks. This question arises because "privateness"[4] in the use of the network drastically reduces the number of subscribers who could legally join the network. Notwithstanding this fact, the answer is positive (Allen, 1988): yes, network externalities do play an important role in diffusion trends of dedicated networks, although their impact over time is different from the case of public use of networks. This assumption can be simply explained by the remark that because of the extremely limited number of people having access to the network, its user-value is expected to be very high for a very limited number of subscribers. In graphic terms, this would mean a logistic curve representing a very high number of adopters in the short run. As Allen (1988) points out, if three networks with a different fixed capacity are taken into account, three different demand curves can be drawn, each of them with a different slope (Figure 2.5). The upward-sloping curve for small networks dedicated to a specific group of people has the largest angle, that is the highest value for the smallest network size. Once we move towards larger numbers of users in the network, as is expected for public networks since they have to reflect the universality rule of a service offered to everybody at the same price, the angle of the demand curve decreases.

The conclusion drawn by Allen from this temporal gap in the distribution of network externality effects is that, taking small networks dedicated to a group of people together with universal networks, the separate critical mass inflection points begin to show a downward sloping demand curve (Figure 2.5). This "long-run average demand curve", as Allen labelled it, reflects a conceptual symmetry with the long-run average cost curve. This conceptual framework is based on the hypothetical assumption that networks gradually shift over time from private to public use, and for this reason it becomes an extremely interesting starting point when problems of interconnection, standardisation and implementation of a single modern public infrastructure need to be taken into account (Section 2.3.3).

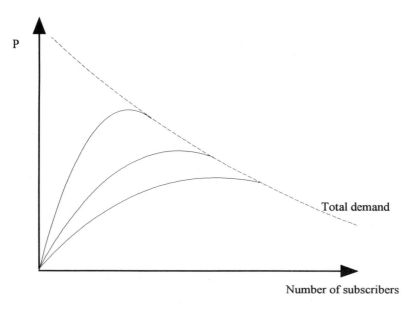

Figure 2.5. The long-run average demand curve

Source: Allen, 1990a

Conceptually, it is clear that the use of a network only to a specific group shifts critical mass in time since its user-value is expected to be very high for a very limited number of subscribers. However, some doubts remain, since:

- empirical evidence has not proved this conceptual issue, because of the difficulty in testing the critical mass point;
- Allen's conceptual framework is built on the assumption of fixed capacity of networks and excludes the possibility of a dynamic technological improvement of the network, which would shift the point of maturity and reinforce network externality effects in the development process. As Hayashi points out on the basis of an empirical analysis on Japanese telephone network

development (Hayashi, 1992), the diffusion process of telephone subscribers seems to find a second point of "take off", a second critical mass, which explains why traffic still increases on the network. This phenomenon, explained by Hayashi on the basis of the recent tendency of dedicated and public networks to interconnect, could also be very well interpreted by the fact that the technological revolution has improved the capacity of public networks by allowing additional traffic on them.

The different effects of network externalities over time also have some strategic implications for the tariff structure, as we will see hereafter.

2.4.2 Tariff structure

Because of the nature of network externalities acting on the price system and on the equilibrium point, their effects have to be taken into account when the tariff structure is considered. The role of network externalities affecting tariff structures is a rather complex issue. However, it is not the purpose of this work to deal with this problem but it is mentioned briefly in this section in order to provide a comprehensive review of the issues linked to network externalities.

As we have already mentioned, network externalities are a typical case of a situation where the Pareto-optimal solution is not reached through spontaneous market forces. As Figure 2.1 above demonstrates, the point where marginal cost is equal to marginal benefit (the intersection point between supply and demand) is no longer valid when network externalities are taken into account. On the market equilibrium, private marginal benefits differs from marginal costs. Some forces, external to the market, have to be imposed in order to restore optimal equilibrium. In particular, the method suggested in the literature (Antonelli, 1989; Hayashi, 1992) would be to provide a government subsidy to telecommunications producers and telecommunications service providers, thus obtaining an increase in the supply (Figure 2.6).

In the case of fixed network capacity, following the rules of the well-known epidemic model the distribution of adopters over time is thought to follow a logistic curve (Griliches, 1959; Mansfield, 1961). Tariff structures could reflect different rates of adoption, i.e. different intentions of adopters to exploit the service. In the first phase of adoption, when the number of adopters is low, the tariff structure has to take into consideration the low level of interest generated by the new infrastructure. In this respect a government subsidy could act as a possible support in the early phase of diffusion. A second phase exists in which the cumulative rate of adopters grows significantly at a positive rate. In this case price can reflect costs, and tariffs can incorporate profit revenues, because the network externality effects act as the major driving force in the diffusion process. At the maturity level, when the diffusion rate is decreasing, congestion on the network is achieved and acts as a negative network externality. In this situation some mechanisms to internalise negative effects into the price system have to be developed and this means that the tariff structure should incorporate the negative externality effects. In the case of congestion, some mechanisms have to be defined in order to internalise the negative effects. The simple question, but at the same time difficult to solve, is: who pays for the negative externality effects? The reply can be twofold, in line with the traditional theory, both responses acting against the principle of universality[5]:

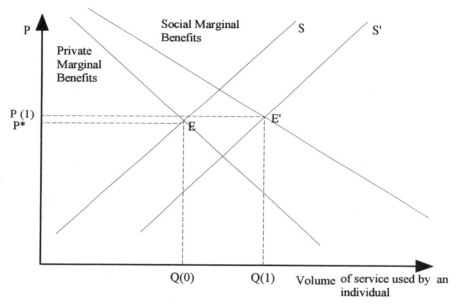

Figure 2.6. Equilibrium in the presence of network externalities

- the already existing subscribers can pay very high tariffs in order to discourage new subscribers to join the network. In other words, the high tariff structure acts as a barrier to entry. In this case the already existing subscribers pay in order to avoid the negative effects of congestion;
- each additional subscriber pays a high entry tariff, which again should act as a disincentive. In this case the new subscribers pay for the "negative effects" they themselves provoke.

Both these suggestions are problematic since they both act against the rule of universality which should be applied to all public goods. In the case of the telephone network, this rule imposes the distribution of the service to everybody without differences in costs among users (Curien and Gensollen, 1992).

2.4.3. Standardisation

The ultimate use of telecommunications technologies is affected by the way they are produced and promoted. A vast amount of literature[6] defines the first phases of diffusion processes of these technologies as being highly dependent on network externality mechanisms, because of the presence of technical interrelatedness. Despite our doubts about whether this phenomenon should be classified as network externalities, this section provides an interpretation of this literature.

In analysing the diffusion of the Qwerty keyboard for typewriters, David (1985) first realised the powerful negative externality effects that standard may generate under certain conditions. It is well known that the keyboard invented by Dvorak is much more appropriate than the Qwerty keyboard developed by Sholes that is now in common use and known as the Qwerty system. Although the Dvorak keyboard is vastly superior to Qwerty, virtually no one trains on Dvorak because there are too few Dvorak machines, and there are virtually no Dvorak machines because there are too few Dvorak typists (David, 1985 and 1992; Liebowitz and Margolis, 1990). The reasons for this inefficient situation have been explained by David (1985) by underlining the success factors of the Qwerty system, namely:

- *technical interrelatedness,* i.e. the need for system compatibility between hardware (the machine) and software (the typists' memory of a particular arrangement of the keys) which makes the value of the hardware highly dependent on the existence of the compatible software;
- *scale economies,* i.e. decreasing cost conditions while increasing the number of people using a particular system;
- *quasi-irreversibility* of investment in specific touch-typing skills.

The case of the Qwerty success in the face of a more efficient technological system is not unique in the history of innovation processes. Postrel (1990) points out the same failure to shift to a better technology in the case of quadraphonic sound, which was unable to replace the stereo system, although its technological features were more advanced. Even this empirical evidence suggests the difficulty of establishing a new technology whose benefits depend on wide adoption, and shows how the problems are magnified by large installed bases of old technology. In both empirical cases, the existence of diffused network externalities, of economies of scale in costs and of sunk costs has acted as a barrier to the shift towards a more efficient technical standard.

Thus, network externalities do not only deeply affect the adoption of a given technological innovation, but also the selection among different competing technologies. Technological rivalry is thus profoundly driven by network externality mechanisms.

The condition under which a superior technical standard is not adopted, labelled by Farrell and Saloner (1985) as a situation of *"excess inertia"*, is thus a type of negative externality and market failure, since each non-adopter of the new standard imposes costs on every other potential user of the new standard. This situation is quite different from the far more common situation where a more efficient standard is invented, but for which the costs of switching are too high to make the switch practicable. In fact, in the case of "excess inertia" the new standard can be clearly superior to the old standard, and the sum of the private costs of switching to the new standard can be less than the sum of the private benefit, and yet the switch still does not occur (Liebowitz and Margolis, 1990).

In other words, adherence to an inferior standard in the presence of a superior one represents a loss of some sort, a cost which is not internalised into the price system. Farrell and Saloner's model is useful because it shows the theoretical existence of a market failure in the presence of an "excess inertia" situation (Farrell and Saloner, 1985). However, Farrell and Saloner (1986) argue that an "excess momentum" may well arise if consumers are forward looking and discount future benefits.

The interesting question is thus how the effects of these consumption externalities can be overcome, or in other words, if there are any methods for avoiding these negative effects. Once again, the question is: *who has to pay for these negative externalities in order to internalise them into the price system thus reaching a welfare state position?*

The creator of a new standard is the natural candidate to internalise the externality, since the creator benefits from the value made available on the market from changing to the superior standard. The greater the gap in performance between the two standards, the greater the profit opportunities are and the more likely the move to an efficient standard will be. There are many ways in which the creator can sustain the costs of transition from one standard to another, and thus internalise the inefficiency:

- discriminating prices between buyers who have already made investments in an old standard and those who have not, in favour of the first group;
- providing subsidies or free training to ensure an adequate supply of operators;
- giving the old standard the possibility of being convertible to the new one.

This last possibility opens up the debate on compatibility among interrelated technologies, i.e. among different networks and different software systems. This issue will be discussed in the next section.

A second possible solution to internalise the negative effects is the payment of a fee by the users of the old technology, which should act as an incentive to move to the more efficient technology.

The example of these technologies opens up the problem of a possible market failure in other technological innovation processes, especially in highly innovative sectors with high sunk costs, with system economies of scale and, moreover, with technical interrelatedness, such as in the telecommunications industry. In this case, the question of a development of a new service on a network with different transmission systems could be hampered by the existence of a very low number of subscribers with a clear intent of not paying for the transition costs from one standard to another.

As in the case of the computer industry, also in the telecommunications sphere standardisation is expected to create negative externality effects because:

- on the demand side, each non-adopter of the new network standard imposes costs on every other potential user of the new network standard;
- on the supply side, the shift to the new network standard is very difficult, if product compatibility is not available.

The logical consequence would be the definition of possible strategic behaviour to avoid these negative effects and to reach a welfare situation. Not many solutions have so far been developed in the literature.

2.4.4. Compatibility and interconnectivity

The existence of different dedicated networks raises the issue of their possible interconnections. For strategic purposes, telecommunications providers can choose to keep their products incompatible, thus reducing the possible size of their networks. Or they can achieve compatibility, either individually (through the choice of technology or by building adapters) or by reaching agreements with their competitors through committees (Tirole, 1988). The issue in this case is thus related to compatibility decisions. Typically, firms can choose whether to manufacture compatible products, and can thus determine whether an individual firm or aggregate market sales are the relevant focus in the evaluation of consumption externalities (Economides, 1992; Katz and Shapiro, 1985). The decision about product compatibility is dependent on many elements, such as:

- the firm size and reputation;

- the market (or "network") size;
- the consumer expectations about a dominant seller;
- the (private or public) incentives to a firm to produce compatible goods or services.

In their theoretical approach to this issue, Katz and Shapiro (1985 and 1986) formalise these elements into a static oligopolistic model and achieve interesting, and realistic, results (Katz and Shapiro, 1985, p. 425):

a) firms with good reputations or large existing networks will tend to be against compatibility, even when welfare is increased by the move to compatibility;
b) in contrast, firms with small networks or weak reputations will tend to favour product compatibility, even in some cases where the social costs of compatibility outweigh the benefits;
c) if consumers expect a seller to be dominant, then consumers will be willing to pay more for the firm's product, and it will, in fact, be dominant.

Most of the literature deals with the compatibility decision in relation to the computer industry, where the so-called software/hardware paradigm represents a typical example of technical interrelatedness in the case of competition. The same problem can be related to the telecommunications sector, where the process towards liberalisation and private provision has generated the problems of compatibility and interconnection among networks. The open questions in this field are the following:

- is the adoption of very technologically advanced networks, such as the integrated service digital network (ISDN) or the integrated broadband network (IBN) limited by the existence of a network standard?
- should privately provided networks be compatible with the public one?
- is compatibility reached through market forces or is public intervention necessary?
- do private networks mutually compete or cooperate?

In the existing literature, these questions are only partially considered and even then are insufficiently answered to obtain some strong orientations at both the industry and the public policy level.

The issue of physical interconnection among networks is becoming a substantial problem, since two contrasting contradictory processes are developing (Noam, 1992):

- the first is caused by the integrative forces of technology which are gradually pushing telecommunications towards a unique public integrated service digital network (ISDN) or the integrated broadband network (IBN);
- the second is generated by social and economic forces of pluralism, which move the network towards a decentralised and segmented federation of sub-networks.

The tension between these two forces is most pronounced at the front where they intersect: the rules governing the interconnection of the multiple sub-networks into the integrated whole. In the coming years, this issue will become more and more crucial, and policy makers have to structure ways in which network interconnection is defined, priced and harmonised at the spatial level. This problem is destined to become one of the major issues for political discussion in the near future.

Until now, however, there has been little attempt to interpret the inter-connectivity issue on the basis of network externalities. Some pioneering contributions have been made by Allen (1988) and Noam (1988), who tried to link the interconnectivity issue with the network externality mechanisms. However, these conceptual contributions lead to opposite results. While Allen suggests a spontaneous process from pluralism to integration of networks, Noam envisages the counter process moving from integration towards pluralism. Even so, both theories have been developed taking into account the concept of network externality.

2.4.5. Economic and spatial symbiosis

The presence of network externalities from the users' side has always been studied as consumption network externalities, since they are related to the interdependence among subscribers in the decision to join the network.

As we have already mentioned in the previous section, there is one aspect that has not yet been analysed[7], which is related to the effects that network externalities produce for the users of the network. In particular, it is our idea that the advantages that a firm receives from using a network are reflected (positively or negatively) in the corporate performance. In the present work, then, network externalities are defined as those technical and pecuniary external economies which generate an increase in input factors' productivity through the advantages obtained by a subscriber joining and using a network, such advantages exerting positive effects on the economic performance of the new subscribers, generating what we call "*economic symbiosis*". By this term we mean the achievement of greater economic performance by exploiting benefits derived from joining and using a network. Once the positive effects generated in the production functions of firms provoke further positive effects within the boundaries of a region, it is possible to argue that the spatial performance may also reflect this (positive or negative) situation, giving rise to what we call the "*spatial symbiosis*" effect. Thus, spatial symbiosis takes place when a set of "networked firms" are present in a specific region and exploit the advantages of being networked.

The above concepts of "economic symbiosis" and "spatial symbiosis" show some similarity to older concepts of "growth poles" (Perroux, 1955) and "growth centres" (Boudeville, 1968), respectively. A growth pole was supposed to operate in an economic network space (without a clear reference to its geographical location), whereas a growth centre was regarded as a geographical concentration of economic activities as a result of agglomeration economies. Both these concepts explained regional and spatial development starting from the consideration of a "key" production unit generating a set of polarisation and development effects, via Keynesian income multiplier effects, direct and indirect input-output effects and spatial advantage effects.

Although our framework suggests that a set of polarisation and development effects arises via the physical linkage of firms, the original causes of the generation of multiplier effects are of a different nature. While the "growth pole" theory defines the reasons for development as the existence of key production units which generate a set of factors attracting economic activity, in the spatial symbiosis approach the cause of generation of these development effects is the actual physical network connection with other firms. This relation, in fact, generates a set of synergies, via exchanges of know-how and information, which improve the productivity of networked firms. At the aggregate level, the greater productivity of firms located in the same region is measured in terms of a better aggregate performance. When the linkage takes place at an inter-regional level, the effects it produces may also be reflected in import and export implications.

The interesting question that stems from this theoretical consideration is, of course, related to the conditions under which this "economic and spatial symbiosis" takes place. The importance of this analysis is quite evident, since the best conditions to exploit the most advanced telecommunications infrastructures may then be highlighted and some policy implications may be drawn. Although some similarities exist with previous theories, a more developed conceptual framework is necessary in this respect, since the subject has hitherto never been tackled at both a conceptual and empirical level. This study is the first methodological, conceptual and normative attempt towards its comprehension.

2.5. Private, club and public goods: the role of network externalities

2.5.1. *Economic definition of different goods: the case of telecommunications networks*

In the previous sections we have presented a typology of network externalities and the implications of their existence on both the production and use of telecommunications technologies. Up to now the concept of network externalities has been related to telecommunications networks in general, without taking into account the different nature of existing networks. In fact, it is our opinion that the degree of implementation of consumption network externalities is related to the *economic nature of the goods with which they are associated*. In other words, the public or private nature of networks and services may encourage or eliminate the existence of consumption network exernalities. The association of network externalities with the economic nature of the technologies involved is something which so far has not been examined in the literature, at least in a systematic way. In this section we present the role that network externalities play when dealing with the different kinds of network.

Telecommunications networks and services have always been regarded as *public goods*, for a number of reasons, and with a number of consequences, namely:

a) until a few years ago the provision of the telephone service has always been developed following the principle of universality. Thus, the use of the service is possible for everyone at the same price;
b) consequent to the first remark, demand for access was not limited by any restrictions; thus access is free to everybody who is interested in the service;
c) also the demand for use was not subject to any particular restrictions, thus even from the point of view of the use, networks could be regarded as public goods.

With respect to the telephone service, then, the term "public good" is used in a rather general way; although subscribers pay for entering the network, their decision to join the network is not restricted by any legal or institutional rule. The possibility to enter the network is equally distributed among potential subscribers, since the price of the service is equal to everybody.

If telecommunications services were simply restricted to the public service, our need to create a typology of telecommunications goods would not exist. In the last decades, however, the public nature of both use and provision of the telecommunications services has changed, and an analysis focusing only on the effects of network externalities in the presence of public goods would be quite limiting and distorted. Two main changes have to be taken into account:

a) telecommunications networks and services have become proprietary and private in their use, and thus they have to be treated and analysed as private goods, in accordance with traditional economic theory;
b) the provision of services and networks is becoming private, driving the telecommunications market towards drastic changes in the institutional rules which govern it, although the degree of liberalisation still differs among countries and for different sub-sectors of the telecommunications industry.

Nowadays there is no simple dichotomy between private and public in the telecommunications network, in that this dichotomy does not cover all "institutional" situations which may be found in the telecommunications domain. If the analysis is carried out from the perspective of the use of telecommunications networks, a range of different goods may be classified under different economic characteristics, which explain the degree of existence of network externalities. Between the two extreme and traditional cases of public and private goods, reflecting respectively a public and private use of the network, we can identify other classes of goods, all having a common feature in a shared use of the network (see Table 2.1). These different goods have been classified on the basis of their degree of *"privateness"* and *"publicness"* with an intermediate "shared area" which includes so-called "heterogeneous exclusive goods", "homogeneous exclusive goods", and the so-called "impure public goods". The economic features distinguishing these different goods are weighted against the baseline of the well-known characteristics of public goods, namely *non-rivalry* and *non-exclusiveness* in the use of the good. In fact, a good is said to be public when the use of it is free to everybody and when the use by one individual of the total amount of it does not interfere with another individual's consumption of it (Samuelson, 1954)[8].

Let us now consider in more detail this intermediate area between purely private and purely public goods. Since Buchanan's paper in 1965 it has been demonstrated that Samuelson's dichotomy between private and public goods was too limited to cover all different kinds of existing goods. It was Buchanan who introduced the concept of "club good", arguing that:

> *"a good is said to be a club good when consumption by one individual reduces potential consumption by others" (Buchanan, 1965; p. 3).*

Table 2.1
Private, club and public goods: the role of network externalities

	PRIVATE USE	SHARED USE			PUBLIC USE
	PRIVATE GOODS	HETEROGENEOUS EXCLUSIVE GOODS	HOMOGENEOUS EXCLUSIVE GOODS	IMPURE PUBLIC GOODS OR CLUB GOODS	PUBLIC GOODS
ECONOMIC FEATURES	Exclusiveness and Rivalry	Exclusiveness, non-rivalry in a heterogeneous group	Exclusiveness, non-rivalry in a homogeneous group	Non-exclusiveness, rivalry	Non-exclusiveness, Non-rivalry
EXAMPLES OF NETWORKS	Intra-corporate networks	Exclusive networks for cooperative relationships	Exclusive networks for competitive relationships	Congested networks	Public networks
ECONOMIC PRINCIPLES	Private marginal costs = Private marginal benefit	Entry marginal costs < entry marginal benefits	Entry marginal costs < entry marginal benefits Marginal Costs = Sum of Priv. Mar. Ben. of members	Social Marg. Costs = Sum of Priv. Marg. Ben.	Social Marg. Costs = Sum of Priv. Marg. Ben.
KINDS OF EXTERNALITIES GENERATED	None	Consumption Extern. Pecuniary Extern. Technical Extern.	Consumption Extern. Pecuniary Extern. Technical Extern.	Neg. Cons. Extern. Neg. Pec. Extern. Neg. Tech. Extern.	Consumption Extern. Pecuniary Extern. Technical Extern.
ECONOMIC CONSEQUENCES	None	Allocative and distributive issues within the group and with non-members	Allocative and distributive issues within the group and with non-members	Allocative and distributive issues among all users	Allocative and distributive issues among all users

In the case of *club goods, or impure public goods*, the missing feature is the non-rivalry in their use. The use of a club good by one individual is dependent on the use of the same good by another individual, although in principle the use of the good is open to everybody.

The opposite case is represented by goods characterised by non-rivalry in its use by different individuals, but with a clear exclusiveness in its use for a defined number of people which we define as exclusive goods[9]. These goods are characterised by some clear economic features, namely (Adams and McCormick, 1987; Cornes and Sandler, 1986; Sandler and Tschirhart, 1980):

a) there must be a voluntary interest in participating in the group, thus the jointly derived utility from membership and consumption of other goods must exceed the utility associated with non-membership status;
b) they oblige a shared use. Sharing, however, may lead to a partial rivalry, causing a detraction in the quality of the service received, where this partial rivalry is dependent on some measure of utilisation;
c) they must be regulated by exclusive mechanisms, whereby users' rates of utilisation can be monitored and non-members can be excluded. Without such a mechanism, there would be no incentives for members to join the club and to pay dues and other fees. The cost of exclusion mechanisms must be less than the benefits gained from allocating the shared good within a group arrangement.

We now define two kinds of goods which possess these above-mentioned features, namely *homogeneous exclusive goods and heterogeneous exclusive goods*. The difference between the two categories of homogeneous and heterogeneous exclusive goods lies in the existence in the second case of well-established property rights among users, a feature which defines these goods as being more similar to private goods, as we will see below.

In the telecommunications context all these different goods in the public and private spectrum are nowadays represented by networks, each having different characteristics in terms of the legal aspects defining their use. A classification of these networks is given below.

Exclusiveness and rivalry are typical of the most private networks, the *intra-corporate (dedicated) networks*, whose use is legally restricted only to a firm, and within its boundaries. Moreover, the use and access to that particular dedicated network becomes immediately unavailable to other users, generating rivalry in its use.

Exclusiveness but non-rivalry in the use characterises exclusive networks for both co-operative and competitive relationships. A clear example of what we mean by *exclusive networks for co-operative relationships* is a network between a firm and its suppliers. In this case the access to the network is closed to some specific firms, whose relationship is a vertical one, and is based on some established and formalised contracts, which define both the object and the terms of the agreement. In this case, property rights and economic interests of the parties involved are quite clearly defined from the beginning.

Exclusive networks for competitive relationships are all those kinds of networks established between firms which have similar economic interests, and which operate in the same market on the basis of strong competitive situations. All networks of industrial associations, such as the SWIFT network, belong to this group. Again, the use of the network is entirely devoted to people belonging to the group, legally excluding others from participating in the network.

Moving towards a higher degree of publicness in the access to the network, we find also the so-called *congested networks*, characterised by public access to the network, but by rivalry in their use. A telephone network, insufficient in its technological capacity to deal with all calls, is an impure public good, since the consumption by one individual reduces potential consumption by others.

The last category is that of *public networks*, where access is free to everybody and their use is unlimited.

2.5.2. Economic principles

As the traditional theory on public and club goods argues, there are very different economic principles which explain the nature of these goods. A private good is said to be private because the marginal costs paid by its user equals the marginal benefits. That is to say, all benefits that a person receives from the use of the good are legitimately his own right because he pays for them entirely. In principle, then, private goods do not contain in their definition any sort of externality mechanisms.

In constrast, an opposite situation is represented by a public good. The concept of public good contains in itself externality mechanisms, since the idea is that the good has to be offered to everybody at the same price, despite the real advantage one can get from it and the use one can make of it. Thus, if private marginal benefits are different from private marginal costs, the familiar problems of market failure and of resource allocation implications arise.

The telephone service has to be offered in accordance with the universality rule, i.e. the telephone service has to be offered to everybody at the same price. In this case, the telephone service becomes a public good, because it confers the same kind of benefits simultaneously to a number of people, and it is impossible to allocate additional costs or benefits to any particular person.

The optimal equilibrium point is thus achieved in the situation where marginal cost (C) is equal to the sum of marginal benefits (V). As Samuelson (1954) pointed out, the optimal welfare state is achieved when:

$$C = V_1 + V_2$$

where V_1 and V_2 represent the marginal benefits of individual 1 and 2, respectively. Because of the public nature of the telephone service, marginal costs cannot be attributed to any particular subscribers without going against the rule of universality. They can only be jointly attributed and, therefore, the jointly incurred marginal cost has to be equated to the algebraic sum of the marginal benefits.

The consequence of the public nature of the telephone service is a situation similar to the existence of externality mechanisms, where the equilibrium point is represented by the intersection between supply and the vertical sum of single demand curves. In this case also, market equilibrium is reached at a point of greater output (see Figure 2.6 above).

The immediate consequence of what has just been said is that in the presence of public networks, all kinds of network externalities are present. The economic consequences are expected to have strong resource allocation implications and distributive issues, related to the distribution of the advantages among users.

In the case of club goods, or impure public goods, the economic principles behind them are the same as those governing public goods, in that non-exclusiveness in the use of the good requires the marginal costs to be attributed to every user of the good, with no linkages with the marginal benefits that a single person receives from it.

With the impure public goods, the problem is opposite to the one of the pure public goods, in that the whole population has to share the disadvantages stemming from the joint use of the network instead of the benefits. Again, all sorts of externality mechanisms (in this case negative externality mechanisms) take place and both resource allocation implications and distributive issues arise.

An interesting case is represented by what we call exclusive goods. These goods are based on the exclusiveness of the use for people belonging to a group, and the exclusiveness is guaranteed by a cost of entry. This class of goods has two different sources of generating externality:

a) as the theory underlines, the cost of the exclusion mechanism must be less than the benefits gained from joining the group. From this definition a first externality mechanism is present since individuals joining an exclusive network do not pay for all the benefits they receive. In this sense, non-paid for advantages are immediately allocated to members of the group, and non-members are inevitably excluded from receiving these advantages;
b) within this group of exclusive goods, externality mechanisms are present, since when the new subscriber joins the network, he does not pay for all the advantages he receives, the network within the group being considered as a

public good. Then, traditional externality mechanisms, due to the joint use of the good, arise.

For exclusive goods, then, there are two sources of economic implications that may arise:

a) the first source is linked to the fact that non-paid for advantages are received by the members of the group, and this provokes resource allocation implications and distributive issues between members and non-members;
b) moreover, the second source stems from the joint use of the good inside the group, where, again as in the case of public goods, resource allocation implications and distributive problems among members of the group arise.

The difference between heterogeneous and homogeneous exclusive goods is based on the fact that the former are characterised by clear allocation of property rights, while the latter are not. The clear allocation of property rights is useful when externality mechanisms have to be overcome, since the definition of the allocation of payments associated with externality effects among members is much easier.

The conclusion of this analysis is that network externalities and the effects they provoke in an economic system are much more evident and likely to occur in the case of public networks. Conversely, they do not exist in the case of intra-corporate networks. Finally, in the case of exclusive networks their effects can be measured both inside the group itself, and between the members of the group and the non-members. The magnitude of their effects will be the subject matter of the empirical analysis of our study.

2.6. The state of the art in the existing literature

Network externalities play a crucial role in many aspects and elements characterising new technologies, affecting both the use and the production of new technologies. As the broad review of the literature shows, there is a variety of problems and discussion elements that come to the fore once network externality effects have to be taken into account. The analysis of the literature on network externalities and on their effects show that there is a growing interest in the production aspects, related to the provision and the development of telecommunications networks. The existing literature on network externalities is very recent, although it already presents a variety of approaches with regard to the different themes (see Table 2.2).

Table 2.2
Taxonomy of existing literature on network externalities

Crucial elements Approaches	DIFFUSION PROCESSES	TARIFF STRUCTURE	STANDARDISATION SAME TECHNOLOGY	STANDARDISATION DIFFERENT TECHNOLOGY	COMPATIBILITY AND INTERCONNECTIVITY
Theoretical and Conceptual	Antonelli, 1989 Antonelli, 1990 Antonelli, 1991 Antonelli, 1992 Allen, 1988 Allen, 1990a Blankart et al, 1992 David, 1992 Griffin, 1982 Linhart et al., 1992 Markus, 1987 Rohlf, 1974 Roson, 1993 Curien et al 1992	Hayashi, 1988 Hayashi, 1989 Hayashi, 1992	David, 1985 Liebovitz et al., 1990	Tirole, 1988 David, 1993 Markus, 1987	Oniki, 1990 Tirole, 1988 Allen, 1992 Antonelli, 1993
Mathematical	Bental et al., 1990	Dhebar and Oren, 1986 Littlechild, 1975 Squire, 1973	Church and Gandal, 1992 Farrell and Saloner, 1985 Farrell and Saloner, 1986	Matutes et al 1988	Noam, 1992 Katz and Shapiro 1985 Katz and Shapiro, 1986 Katz and Shapiro, 1992 Economides 1989 Economides 1992 Amiel et al 1987 Curien et al. 1987
Empirical	Antonelli, 1990 Antonelli, 1991 Cabral et al, 1991 Antonelli et al 1989	Hayashi, 1992	David, 1985 Postrel, 1990		
Policy oriented	Antonelli, 1989			Cowan and Waverman, 1971	Noam, 1987 Noam, 1988 Noam, 1992 Oniki, 1990

The analysis of the implications of network externalities on the tariff structure is still very limited, so that in this case a combination of this literature and that of the domain of public finance and public economics seems necessary.

In the case of standardisation, David's paper on the Qwerty phenomenon has opened up the debate in this field, and some strong contributions have been provided from a mathematical point of view in order to improve this analysis. Yet, these are only a few attempts to find possible solutions to avoid the negative effects provoked by the persisting old standard. Some mechanisms can be put in place from an economic point of view, in order to internalise these negative effects into the price system, thus achieving an optimal welfare solution.

Concerning the interconnectivity problem, this is destined to become a major policy issue in the telecommunications field in the next decade.

The migration strategy from the present networks towards ISDN has yet to be established, and the present debate is still rather limited to very few contributions. Many studies have been conducted with a mathematical approach, but still empirical evidence is lacking, as well as contributions to the policy-oriented side.

The most developed and analysed research subject in the field of network externalities is related to the diffusion processes of new technologies, i.e. on consumption network externalities. Many interesting contributions have been written, concerning theoretical, mathematical, empirical and even policy-oriented approaches. All these contributions deal with the effects of network externality on the diffusion of new technologies over time, with network externalities being one of the most important reasons for their adoption. However, no attempt has yet been made to discover the effects network externalities produce on the performance of users, which we classified in Section 2.3 as the effects stemming from production network externalities on the users' side.

The measurement of what the presence of network externalities means at the level of industrial and spatial performance is something which has hitherto never been discussed.

2.7. The contribution of the present work: research issues

In this chapter, the concept of network externality has been discussed at length. It has been demonstrated how this concept has great influence on many aspects of the new technology development. It is only recently that its importance has been recognised and discussed in a still limited number of contributions. Although at present recognised as one of the main reasons explaining the technological adoption of interconnected technologies, in the case of the diffusion of new technologies the concept of network externality still lacks of a solid theory, especially when the effects of the existence of network externalities on the demand side are taken into account.

A still unclear definition of the concept itself reflects the absence of a well-developed literature in this field. Especially, there is a habit to label many different phenomena under the same heading of "network externalities". There has never been an attempt to classify all possible network externalities within one typology, to help conceptualise the economic consequences and effects that their existence produces, and the way in which these effects may be solved.

While it is very clear in the literature that network externalities are the reasons for entering a network, as they have a direct impact on the utility function of each user, an analysis of what the effects of their existence means at the level of industrial and spatial productivity, and thus performance, is still lacking. The aim of the present work is to provide a contribution in this still unexplored area. In particular, our objective is to develop a conceptual framework of analysis arguing that the existence of network externalities exerts a clear effect on the economic performance of firms and regions, which we have already labelled as an "economic and spatial symbiosis".

Moreover, this conceptual focus will be tested at an empirical level, in order to prove its validity. Such an effort has to our knowledge never been tried before. Our idea is that a better economic performance may originate from the effect network externalities produce. Thus, given this assumption, the definition of network externalities in this work has been described as *"those positive advantages (not entirely paid for) that a new subscriber receives (or generates) from joining and using a network, which produce effects on the economic performance of adopters"*.

Consequently, a *first basic research issue* of the present study will be:

1. *to determine the importance which network externalities have on the decision of new adopters to join the network. In other words, the first aim of the study will be to determine to what degree network externalities play a fundamental role in the diffusion of new and modern telecommunications networks and services.*

Our idea is that network externalities represent one of the most important reasons for the adoption of these open networks. Although this has already been recognised by many studies, its empirical validity is something that so far has not been satisfactorily studied, at both a micro- and macro-level. Our idea is to test this hypothesis in order to define whether network externalities influence particular firms and are induced by seedbed areas. In this case we will identify which categories of firms and regions are more inclined to be influenced by network externalities in their adoption decision process.

Some methodological problems have to be faced in this respect. No rigorous attempts have so far been developed in order to measure this phenomenon, therefore a framework for analysis will be discussed in one of the next chapters. Moreover, in fact, very little previous work has been done to investigate the effects of these network externalities. In this study, the idea is that network externalities are much more evident in public networks and also in exclusive networks, since the use of the network by their respective subscribers generates advantages. Private networks do not generate any kind of network externality effects, since the marginal costs equal the marginal benefits for the new subscriber.

The *second crucial research issue* is thus related to the effects that network externalities generate for users of public or exclusive networks. The research issue will be:

2. *to see whether network externalities provoke an increase in the productivity of firms and regions. More precisely, this second issue will try to answer the following questions:*

   2a. Can network externality effects be measured at the firm's performance level?

   2b. Can network externality effects be measured at the spatial performance level? In other words, have regions the possibility to gain from these network externalities?

A *third research issue* will be:

3. *to define under which micro- and macro-conditions network externalities play a role in the performance of firms and regions. Clearly, network externalities do not always generate positive effects on regional performance. This would be true if space were homogeneous, but the economic discrepancies among regions determine differences in the capacity of a region to exploit these technologies. In other words, we expect regions to exploit network externalities with different intensities and thus obtain different results.*

The existence of network externalities in the adoption of networks does not necessarily mean advantages for firms and regions. Some other economic features of an area may also contribute to the exploitation of these technologies. The literature has already pinpointed these features as being psychological, cultural and social variables which define the level of education, the willingness to take risk and the attitude to change.

The most important aim highlighted in this research issue is to find out under which conditions winners and losers will emerge. The conditions defining who is the "loser" and who is the "winner" in a network is strictly dependent on the capacity of a region (or a firm) to exploit these new technologies. As Cappellin and Nijkamp (1990) suggest:

> *"It is clear that the emergent new technology offers a window of opportunities for regional revitalisation. Whether such opportunities will be grasped is a matter of local entrepreneurial spirit and policy creativeness" (Cappellin and Nijkamp; 1990, p. 5).*

In other words, the actual exploitation of "network externalities" is very much related to the organisational capability of the local environment. The capacity to accept the new technology and find the right blend between technological potentialities and industrial practices is of crucial importance if local economies do not want to play the role of "loser" in the network which would mean the loss of advantages generated by positive network externalities.

The role of technological change in regional development is far from being a new research issue. Therefore, a review of what has already been said on this subject is a necessary starting point if we want to analyse the role of new network externality effects at a spatial level. Thus, in the next chapter we will provide a concise review of the literature on the spatial aspects of technological change, in order to prepare the ground for our theoretical and empirical framework, which will be presented in the remaining chapters of the book.

Notes

1. See, among others, Allen, 1988 and 1990a; Antonelli, 1989, 1990 and 1992; Bental and Spiegel, 1990; Cabral and Leite, 1989; David, 1985 and 1992; Hayashi, 1992; Katz and Shapiro, 1985; Markus, 1989.
2. For a review of the concept of externality see Coase, 1960; Cowen, 1988 and Mishan, 1971.
3. The externality curve is very difficult to construct, as all good "Public Economics" textbooks teach (see for example, Brosio, 1986). It is in fact very difficult to evaluate the social benefits with an objective measure. For the purpose of our analysis it is nevertheless very useful to provide a measure, albeit in qualitative terms, of the externality effects on the traditional equilibrium point.

4. In this context by "privateness" we mean the use of a network restricted to a specific group of people (or firms), and not to a single individual (or firm).
5. The principle of universality guarantees that the use of the service is possible for everybody at the same price, despite the costs of production, which can differ from one region (or subscriber) to another. Cross-subsidy mechanisms among different services or categories of users exist, which allow the universality principle to take place.
6. See, among others, Allen, 1992; Amiel and Rochet, 1987; David, 1993; Katz and Shapiro, 1985 and 1986; Economides, 1989 and 1992; Farrell and Saloner, 1985 and 1986; Matutes and Regibeau, 1988.
7. Antonelli (1993) has recently provided theoretical definitions for the different kinds of production network externalities.
8. Following Samuelson's definition of a public good, the telephone service should not be treated as a public good, since subscribers have to pay for it. In our study, however, the term "non-exclusiveness" is used in a more general way, and means a situation in which everybody is free to choose whether he wants the service or not, without any legal and institutional rule influencing his choice. The "non-rivalry" in the use is present also in our definition of a public good.
9. These goods have been recently called in the literature "club goods". However, in this study we use the original definition of club goods given by Buchanan, while we label those goods having the characteristics of non-rivalry and exclusiveness as eclusive goods.

3 Regional aspects of technological change

3.1. Introduction

Until the 1980s, technology and innovation were widely overlooked in the explanation of regional development and regional growth differentials in advanced industrial nations. This situation changed during the 1980s, when technological changes and innovation diffusion became a central subject in regional studies, thus superseding the neo-classical explanation of technology as "manna from heaven".

The recent keen interest in technology and innovation has been mainly brought about by the inadequacy of the standard neo-classical theory to explain the rapid decline of economic growth rates in the 1970s and the 1980s. The interpretation of technology as a "residual" part of an economic growth rate, that cannot be attributed to growth in labour and capital productivity, provided a more or less sufficient explanation for economic development in periods of stable and constant growth, as was the case during the 1950s and 1960s. Regional growth rate differentials were simply attributed to differences in regional rates of technological progress, leaving unexplained the differences in regional production functions (Malecki and Varaija, 1986; Kamann and Nijkamp, 1991).

The inadequacy of such an approach became evident at the beginning of the 1980s, after a period of decline in advanced regions and of industrialised national economies. In those years economic analysis focused on those concepts and models which could explain long-term fluctuating patterns more adequately. The current focus on technology as the driving force of economic growth is not only typical of regional economic analyses, but is also the result of a general "upswing" in the interest in the role of technological change in economic theory. It was in fact during the 1980s that the concept of technological trajectories was elaborated. This concept referred to the long-term development associated with the physical specificities of each technology, and of the automatic processes of incremental innovations which determine the upgrading of technological performance. The idea of technological trajectories gave rise to the so-called "evolutionary" approach to economic growth (Dosi, 1982; Dosi et al., 1988; Freeman et al., 1982) (see Chapter 1).

The acceptance of the strategic role played by technology in economic development during the 1980s is witnessed by the European Community's launch of a series of extensive programmes in Research and Technology Development (RACE, ESPRIT, BRITE, STAR, etc.), in order to promote the creation of a new European-wide space in science and technology (Quévit, 1990).

Moreover, at the beginning of the 1980s, the traditional regional economic policy of financial incentives to encourage the physical transfer of economic activities from industrialised to backward regions showed its inefficiency as an instrument to decrease regional economic disparities. During the 1980s, the aims of regional policies changed drastically. In these years because of the strong interdependence between innovative activities and economic growth, the objectives of regional policies shifted to the promotion of new technologies in traditional and new firms, with the aim of strengthening regional economies (Ciciotti, 1984; Davelaar, 1991; Ewers and Wettman, 1980).

In accordance with theories of industrial economics, the spatial generation, diffusion and effects of new technologies have become an indispensable point of departure to explain regional economic growth and disparities.

This approach can be included in the general framework of "*endogenous development*" theories. Since the 1970s, in fact, regional development and regional growth disparities have been studied under the assumption that endogenous resources are decisive in fuelling economic development, these internal resources being the entrepreneurial capabilities, local technical and scientific know-how, flexibility of production and traditional location factors (Cappellin, 1983). The so-called "*development from below*" approach to regional development studies seems to be the right one during years of reduced spatial mobility of both labour and capital (Stöhr and Tödtling, 1977), since the "exogenous" approach explanation of regional development is mainly based on inter-regional movement of capital and labour.

Since the 1980s productivity, continuing innovation and technological changes have been considered as the main local catalysts for regional economic development, thus abandoning the standard neo-classical concept of technology as a given resource and explaining both the development mechanisms of single regions and the general evolution of inter-regional disparities.

As a consequence, a series of theories have been developed concerned specifically with the spatial aspects of technological change. Some of them focus attention on the *region* as the subject of study, trying to capture the level of innovativeness of each single region, and particularly looking for those local innovative potentialities which explain local competitiveness. Others, on the contrary, approach the problem by concentrating on *technology* as the subject of study and analysing its spatial development trajectory.

The approaches to the explanation of both regional development and regional disparities are taking place at both a micro- and a macro-level. The micro-approach is the most common and involves the analysis of the structural and sectoral components of a region, looking at the industrial mix between innovative and traditional sectors located in the region. The "micro" cosmos of the firm with its innovativeness influences the regional performance. On the other hand the macro-approach relates to the different regional endowment of specific resources, where the analysis focuses on local economic resources rather than on the performance of firms.

Generally speaking, the logic underpinning the relationship between technology and regional development is a circular one. New innovative activities are in fact expected to follow a spatial adoption pattern, this adoption process in its turn giving rise to new spatial structures, through the location of new more advanced firms or through the restructuring of already existing firms which then become more efficient and competitive. From this new and more efficient spatial structure a greater regional performance is expected which again stimulates, via greater innovative potentialities, new technological changes (Figure 3.1). Thus, technological change creates new spatial structures, and radical changes in technology - i.e. the emergence of new technology systems - are expected to lead to structural, though discontinuous, changes in the spatial economy (Figure 3.2). These structural changes are inevitably expected to generate innovative activities in the long run because of the change in the structural and sectoral regional components in favour of more innovative sectors.

This circular reasoning, though very simple in its presentation, contains some key elements that have to be analysed in order to be able to understand the spatial trajectories and effects of the technological change, namely the identification of those elements which explain:

a) how technology develops in a spatial context;
b) what are the effects of technological diffusion on regional performance.

The first of these key elements relates to the sphere of adoption, to the regional generation and diffusion of new technologies, to the spatial dynamics of technological changes, and to the definition of spatial determinants facilitating or preventing a local economic system from being able to innovate. The second question refers to the effects generated by new technologies on regional performance and disparities.

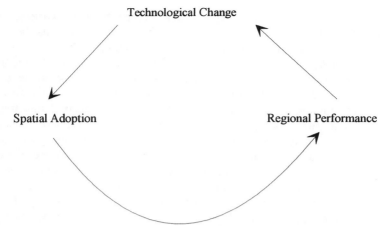

Figure 3.1. Linkage between technological change and regional performance

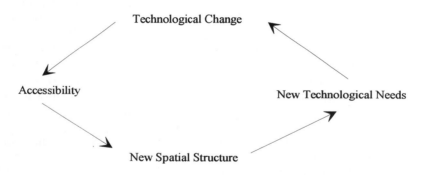

Figure 3.2 Linkage between technological change and new spatial structure

The attempt to answer to these questions is far from new. Since the late 1970s, many theories have been elaborated on both the "*adoption*" and "*performance*" aspects of spatial patterns of new technology. The aim of this chapter is to provide a structured and critical review of these theories, keeping a distinction between the "regional adoption" theories (Section 3.2) and the "regional performance" theories (Section 3.3). Table 3.1 provides a summary of historical approaches to technological diffusion and regional performance.

Table 3.1
Typology of approaches to regional aspects of technological change

INNOVATION PHASES APPROACHES	SPATIAL ADOPTION	SPATIAL PERFORMANCE
MICRO (Firm level)	* Presence of adjustment costs (Sherer, 1980; Camagni, 1985) * Relative profitability (Camagni, 1985) * Availability of information	* Productivity models (Camagni and Cappellin, 1984)
MACRO (Regional level)	* The single technology model (Hägerstrand, 1967; Griliches, 1957; Mansfield 1961; Camagni, 1985; Capello, 1988) * Innovation potential model (Oakey et al, 1980; Malecki 1979, etc.) *Attractiveness potential models (Townroe, 1990) * Epidemic models (Norton and Rees; 1979; Vernon, 1966; Davelaar and Nijkamp, 1990)	* General development theories * Regional development potential theories Biehl, 1986; Blum, 1982; Bruinsma et al, 1989) * Development theories based on transport systems

3.2. Spatial adoption of new technologies

The first step towards the interpretation of the role of technology in regional development is the analysis of the spatial dynamics of technological change.

The introduction of the spatial dimension in studies of innovation diffusion does not mean the introduction of a further variable in an already very complex framework, but it plays a fundamental role in the explanation of some crucial aspects of the present (and future) diffusion processes. Moreover, the interest in the analysis of space in the innovation diffusion processes is something that was already present to some extent in various studies of industrial economists. Evidence of this phenomenon can be found in the concepts of "selection environment" introduced by Nelson and Winter (Nelson and Winter, 1977 and 1982), of "industrial atmosphere" of Marshall (Marshall, 1919) and of product life cycle (Vernon, 1957) first developed in urban economic contexts (Camagni, 1985; Kamann, 1986). In other words, space matters, and innovation diffusion processes can only be fully explained by also taking into account the spatial context in which they are developed.

A pioneering study in this respect was the analysis undertaken by the geographer Hägerstrand (1967), who interpreted the diffusion pattern of a new technology as being dependent on the communication and information channels of a specific area. According to his view, the spatial dynamics of a new technology were driven by its capacity to receive information through formal and informal channels, so that learning and communication processes were the main explanation for a spatially differentiated distribution of innovative activities.

On the basis of the generally accepted S-shaped function representing the temporal diffusion process (Andersson and Johansson, 1984; Davies, 1979; Griliches, 1957; Mansfield, 1961 and 1968; Metcalfe, 1981), Hägerstrand tried to associate every temporal phase of development with a spatial pattern of diffusion. In particular, Hägerstrand was speaking about three main spatial effects of communication and learning diffusion channels, i.e. the *filtering down processes,* the *neighbourhood effects,* and the *inter-city linkages,* which are associated with the different temporal phases as follows:

- in the first adoption phase, technology develops spatially following the filtering down process through the urban hierarchy;
- in the diffusion phase, spatial technological patterns are characterised by both the filtering down effects and the epidemic, or neighbourhood, effects;
- on reaching saturation, the spatial development follows a random pattern.

This study was further developed by Pred, who demonstrated that the flows of specialised information are spatially biased towards the larger metropolitan complexes (Pred, 1977; Pred and Tornqvist, 1981).

In these analyses, it is implicitly assumed that every potential adopter has an equal opportunity to adopt and that spatial variations in adoption rates depend on spatially biased information flows. The limitation of this approach lies in the fact that only contact probabilities and information exchanges are considered to be relevant in the innovation process, this probability being simply a reverse function of distance in the three above-mentioned channels (Camagni, 1985). In Hägerstrand's model information implies adoption, and when he tries to include the different receptiveness of the potential adopters in his model, he refers again to a measure of information intensity:

> "*a person becomes more and more inclined to accept an innovation the more often he comes into contact with other persons who have already accepted it*" *(Hägerstrand, 1967; p. 264).*

What is really missing in Hägerstrand's contribution is a more realistic analytical explanation of the spatial variations in technological adoption rates, which can supersede the simple interpretation based merely on spatially biased information flows. The element which underlines the inadequacy of Hägerstrand's approach is the spatial element itself, which needs to incorporate the different receptiveness of potential adopters and thus justify the different spatial trajectories of technological adoption (Brown, 1981; Camagni and Pattarozzi, 1984; Capello, 1988; Davelaar, 1991).

Therefore, nowadays the simple explanation given by Hägerstrand is less acceptable in spatial studies. From the new perspective, two main types of studies can be distinguished, which place much more emphasis on the spatial determinants of innovation processes, namely (Momigliano, 1984):

- a first body of literature focusing attention on the spatial processes of R&D and innovative activities, thus looking precisely at the *spatial innovation potentials*;
- a second body of literature dealing with the inter- and intra-regional diffusion of innovation. Moving on from the criticism on Hägerstrand's explanation of spatial variations in adoption rates, this group of studies further develops the theme of *spatial adoption potentials.*

In the following sections, a review of the literature will be provided for both the spatial innovation and adoption potentials in Sections 3.2.1 and 3.2.2, respectively.

3.2.1. Spatial innovation potentials theories

The basic idea of this group of studies is the analysis of the potentialities of each local area to innovate. The region thus becomes the central theme in a great number of empirical studies where the aim is to obtain regional indices of innovation inputs and outputs. As Davelaar points out very clearly (Davelaar, 1991), these studies are much more oriented towards the analysis of the spatial generation of innovation activities, rather than towards the diffusion processes, where the difference between "generation" and "diffusion" of innovation is related to the temporal subdivision of development phases (Stoneman, 1983). In these studies the regional indigenous potential is at the basis of the different levels of regional innovative activities and at the same time it explains the different qualitative types of innovations existing in a specific area.

A great number of mainly empirical studies have been undertaken, with a bias towards the analysis of innovative capacities in high-tech firms rather than in traditional industries, because of the greater expectation of innovative activities in new sectors (Davelaar, 1991).

As far as innovative input studies are concerned, the most common indicator is the intensity of R&D activities in an area, which has been studied in aggregate analyses by looking at the total R&D activities in terms of the dispersion of laboratories or R&D employees (Bushwell and Lewis, 1970; Clark, 1972; Malecki, 1979; Howells, 1984 and 1990; Quévit, 1990; Thwaites and Alderman, 1990); in analyses on the spatial dispersion of R&D activities in large "high-tech" (multi-located) firms (Malecki, 1980; Howells, 1984 and 1990; Gibbs and Thwaites, 1985); and in analyses on the spatial dispersion of R&D activities in small and medium sized "high-tech" firms[1] (Oakey, 1984). Although relevant only for the UK, Oakey et al. (1980) measure the innovative potentialities of British regions through a rather unusual indicator of innovation, the number of awards given to British industrial firms for achieving either an increased volume of exports or new technological innovations (Camagni and Cappellin, 1984; Molle, 1984).

A second approach to the measurement of innovative potentials of regions is by focusing on regional innovative output distribution. A traditional indicator of such a variable is the number of patents granted to firms located in a specific area (Boitani and Ciciotti, 1990; Ewers et al., 1979; Pavitt, 1984).

The stimulating and satisfying element of all these studies is that two general common conclusions are reached. The first common result in these analyses is that innovation activity has a *natural tendency towards spatial concentration*. This is true considering measures of both input and output innovative activity. Central regions are those having the greatest number of R&D activities, as well as of patents. This first general result can be explained by the fact that concentrated locations allow for easier exploitation of technological and scientific know-how developed in research centres and universities, easier access to tacit information, a wider mobility of skilled labour force and a wider development of advanced services. This also explains why R&D activity location can be summarised in three typologies, namely a) large metropolitan agglomerations; b) cities and industrial districts; c) specialised "valleys" or "corridors", such as "science parks" (Momigliano, 1984). In all these three locations the proximity of economic and scientific operators is a necessary condition for the creation of tacit information flows, of a host of cross-fertilisation processes, of synergies among research and production units, and of spin-off effects through skilled labour force mobility (Goddard, 1980). These facts and processes bring about very powerful polarisation effects in research activities, innovation and advanced industries, different in nature but similar in their spatial consequences to Myrdal's well-known "backwash" and "cumulative causation" effects which explained the spatial concentration tendencies of the first part of this century (Camagni and Rabellotti, 1990).

However, nothing new has been said above: since Marshall's "urban agglomeration economies" and the early studies of Walter Isard, the concept of urban synergy and of urban attraction of economic activities has pervaded regional studies. What is new in this respect is the current emphasis given to this concentration phenomenon, thanks to the ever increasing importance of a typically urban and strategic production factor, namely information.

An extrapolation of this polarisation effect can explain the second result of the analyses on spatial variants of innovative potentials, namely the fact that innovative activities are characterised by a *cumulative nature*, justified by external economies, by spin-off effects of skilled labour force and technological and organisational know-how which taken together develop locally a process of cross-fertilisation of technological innovation (Goddard, 1985; Goddard and Thwaites, 1980). The model of *"rupture/filiation"* (break and continuity) developed by Philippe Aydalot (1986) summarises the concept of cumulative process: if innovation always implies a break and a bifurcation with respect to past experience, nevertheless it finds its roots in the history of a local community, as it feeds on some pre-existing know-how, synergy or tradition.

Despite the very homogeneous results of these studies and their widespread range, covering all the most developed European Countries, this approach to the analysis of spatial innovative potentials can nevertheless be criticised with respect to some of its hypotheses and methodology. In particular, criticisms concern a) the validity of the specific indicators used for measuring innovative activities, such criticism being raised also in industrial economic analyses; b) the capacity of this approach to clarify spatial elements of technological change.

Concerning the validity of the indices used, the main doubt is related to the fact that knowledge is only generated in R&D laboratories. Analyses exploring sources and causes of technological change have brought about several other information producing channels, such as universities, government agencies and non-profit centres. Moreover, firms may acquire technology knowledge (totally or, more frequently, partially) through external channels, via (Belussi, 1990):

- "reverse engineering" techniques (Camagni, 1986; Pavitt, 1984);
- licences purchase;
- technology embodied in new machinery;
- all forms of learning (Camagni and Capello, 1991; Gambardella, 1985);
- relationships with customers, suppliers and subcontractors (Antonelli and Gottardi, 1988);

- recruitment and training activities of the workforce (Freeman, 1987).

Also for the output index expressed in terms of the number of patents, some doubts remain concerning the capacity of this index to incorporate regional innovation potential:

> *"There remains that doubt which arises from the deficiencies of patents as indicators for innovation performance: not all patents lead to marketable innovations, and by no means all innovations are based on patented innovations" (Ewers and Wettman, 1980, p. 167, quoted in Davelaar, 1991).*

A major criticism of this approach is related to its appropriateness in explaining spatial variations in innovative activities. In fact technology is not only the result of endogenous innovative capacities, but can stem from some other endogenous elements not directly related to the innovative process. In fact, technology can come to a region along five different pathways (Townroe, 1990):

- *new plants and equipment*. As industrial and commercial firms purchase new plants and equipment incorporating new capabilities, they acquire new technology.
- *Migrant company investment*. A new branch plant or a company transferring operations into a city region will mean new technology, with some degree of spillover into the innovation capacity of the region.
- *Intra-company transfers*. A firm with establishments in several different cities will be transferring skills and know-how between plants.
- *Local research and development expenditure*. An investment may be designed to generate technological advance in a company. This may generate benefit to other local companies, particularly customers and suppliers.
- *Spin-off*, with two meanings: a) the external economy obtains advantages because one company benefits technologically from a neighbour's advance in technology; b) one company obtains advantages from another because it learns from it, through local information flows.

These pathways of innovation activities witness the fact that the innovative capacity of a region, as for a firm, cannot be measured only through its "direct" innovative capacity. An indirect capacity should also be taken into consideration, i.e. a capacity related mainly to the attractiveness of the region for external innovative inputs. We do not refer here to the old dichotomy between endogenous and exogenous elements for regional development. Always within the definition of endogenous local capacities to innovate, it is necessary to take into account a distinction between spatial innovative potential and spatial potential of attractiveness of innovative activities. Both of these variables have to be analysed, because of their complementary role in the explanation of spatial innovation capabilities.

The spatial potential of attractiveness can be measured by the number of new firms, as well as the number of new establishments from multi-location firms, with particular new capital endowment, coming into the area. Thus, the exploitation of the traditional location factors, such as easy accessibility to the area, infrastructure endowment, competitive production costs, availability of specific labour force, can turn out to be a complementary element to innovative potential for measuring regional competitiveness.

Recent works have stressed the locational attractiveness of certain areas for new high-tech firms. The reasons justifying these studies are that a) high-tech industries have a very high innovation rate; b) since high-tech industries display certain particular features, which are different from traditional industries, they are expected to follow different locational patterns. In particular, despite a certain kind of apologetic literature claiming that high-tech industries are footloose as far as traditional location factors are concerned, a certain pattern of location is in fact common to these new firms (Camagni and Rabellotti, 1988; Decoster and Tabariés, 1986).

While it is undoubtedly true that the importance and combination of location factors are not the same for each production, technology and firm size; the following factors are the ones most commonly suggested in the economic literature for explaining the location patterns of high-tech firms (Stöhr, 1986):

a) skilled labour-force;
b) amenities, like housing supply and educational and cultural infrastructure endowment;
c) infrastructure and space to grow;
d) university and public research institutes;
e) business climate.

Once again, these location factors are usually found in central areas, where in fact high-tech firms locate the innovative part of their activities, leaving the most operative and standardised work to more peripheral areas. On the basis of their analysis of high-tech firms in the Milan area, Camagni and Rabellotti (1988) come to the conclusion that high-tech firms in Milan cluster around a specific area, the North-Eastern part of the town; this cluster was initially rather coincidental, but during the last decade it turned out to be a very favourable location, thanks also to the very easy access to the airport. The spontaneous choice at the beginning has been later followed by a rational decision-making process, again reflecting the fact that traditional location factors are rather important.

Thus, innovation generation stems from both a local innovation potential of a certain area, and those elements that render this area an attraction for introducing innovative potentialities. In the above-mentioned taxonomy developed by Townroe, both new plants and equipment and intra-company transfers depend on the innovative attraction of the area. Focusing on just the internal direct innovative capacities of a region would be rather a misleading approach to measure the generation potential of a local economy.

At the end of the 1980s and beginning of the 1990s the concept of "*milieu innovateur*" has been developed (Camagni, 1991b; Maillat and Perrin, 1992; Maillat, Perrin and Senn, 1993). This approach refers again to the study of the region as the main focus of the analysis, looking precisely for those elements and economic and industrial characteristics in the local economy that can justify the innovativeness of an area in a *dynamic perspective* (Porter, 1990).

This theory is a modern version of the concept of "industrial district" or "systems area" developed in the 1970s in the framework of the endogenous growth approach (Bagnasco, 1977; Becattini, 1987; Brusco, 1975; Fuà and Zacchia, 1983; Garofoli, 1981). What is new and innovative in this theory, is the importance given to technological change and innovation processes as one of the elements which explain the relatively positive performance of the area, rather than only the efficiency factors of local economies. This approach is thus much more related to dynamic factors, such as technology creation capability, fast reaction capability to shift resources from declining production to new production utilising the same fundamental know-how and the capability of regeneration and restructuring of the local economy once hit by external turbulence (Camagni, 1991c).

This theory is already oriented towards the explanation of the development of specific local areas through the definition of those dynamic elements (dynamic efficiency elements, proximity economies, synergy elements) which determine the mechanisms and the forces that lead to better local economic performance.

3.2.2. Spatial adoption potentials theories

A second group of theories focuses the attention on the *spatial adoption processes*, by placing the technology, and no longer the region, at the centre of the analysis. Different from the previous analyses, which consider technology as an endogenous phenomenon that can be manipulated by means of input efforts devoted to it (Davelaar, 1991), these theories concentrate on the spatial dispersion of specific technologies. They place emphasis on a given technology and how it develops over space.

These analyses overcome the above-mentioned criticism of the theory of Hägerstrand (i.e. concerning his unrealistic hypothesis that information necessarily means adoption) and instead place emphasis on the spatial determinants as the elements influencing the receptiveness of a region to adopt a new technology (Figure 3.3). In fact, a possible way of overcoming Hägerstrand's limiting assumption, without abandoning the interesting results already present in the model, is to express the spatial element as *economic distance* rather than *physical distance* and to incorporate in the analysis the economic differences of the regions. In fact, these economic differences represent the reasons for the different degrees of spatial innovative receptivity (Griliches, 1957; Mansfield, 1961; Camagni, 1985).

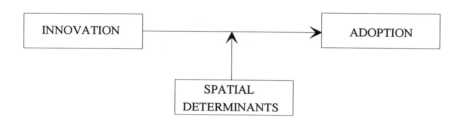

Figure 3.3 Linkage between innovation and adoption

Based on this logic, the pioneering studies of Griliches (1957) and Mansfield (1961) incorporate into the analysis of the spatial diffusion of new technologies the element of *economic distance* between regions, developing the model in two phases. In the first phase, the aim is to define the temporal development of a specific technology, accepting the logistic curve as one way of studying the innovation process over time (Pasini, 1959). Analytically, the logistic curve reads as:

$$Y = \frac{K}{[1+\exp(-a-bt)]}$$

where:

- Y represents the adoption density,
- a represents the moment in time when the first adoption takes place,
- b is the diffusion coefficient and measures the speed of the process,
- K is the application potential, that is the maximum percentage of potential adopters to which the logistic curve tends asymptotically.

The advantage of the logistic curve is its easy parameter estimation. A simple representation in logarithmic terms of the above-mentioned function leads to a linear function such as:

$$\ln\left(\frac{Y}{K}-1\right) = a + bt$$

where the linearity facilitates the econometric estimation of parameters. The extrapolated logistic curve represents the whole diffusion process in time for the same technology in different regions.

In the second phase, the interpretation of the spatial differences in the diffusion process takes place due to inter-regional cross-section regression analysis between the logistical parameters "a", "b" and "K", and the variables representing the structural and spatial-economic characteristics of the region (X, Y, Z....), thus explaining the spatial dispersion of the technology on the basis of the spatial determinants. This reads as:

$$a = f(X, Y; Z, ...)$$
$$b = f(X, Y; Z, ...)$$
$$K = f(X, Y; Z, ...)$$

This two-stage procedure allows for both the estimation of the future potential development of a technology and also the definition of the fundamental spatial elements determining such a spatial bias in the diffusion of a certain technology.

Griliches (1957) and Mansfield (1961) applied this method to the study of the diffusion of new technologies in the agricultural industry and in some manufacturing sectors, respectively. Griliches' work is in fact an attempt to understand the differences in the percentage of the total corn acreage planted with hybrid seeds, by states and by years, thus explaining the way in which technological change is generated and diffused in US agriculture. Alternatively, Mansfield focuses on the diffusion of 12 particular innovations spread from enterprise to enterprise in four manufacturing industries (bituminous coal, iron and steel, brewing and railroads).

On the basis of these two pioneering studies, Capello (1988) applied the same method for the telephone service and its diffusion in the 20 Italian regions. The estimated logistic curve fitted the real data on the diffusion of the telephone service[2] particularly well. Moreover, the regression analyses undertaken gave successful results, showing that:

- the saturation level (parameter K) is highly dependent on the industrial performance of a region and on the education level;

- the moment in time when the first adoption takes place (parameter a) depends on the urban structure of the region, reflecting once more the centripetal process followed by the innovation diffusion;
- the diffusion speed (parameter b) turns out to depend on the economic and industrial level of the region. The higher the economic and industrial level of the region, the quicker the introduction of the telephone service.

These results stress the idea that technological innovation follows *centripetal processes* and starts first in regions with a higher potential demand density.

Despite the positive results and the original idea of considering innovation diffusion as a *continuing process*, this model related to the logistic function has some limits. One limiting hypothesis of the model is that technology does not evolve over time (Davies, 1979; Stoneman, 1986). As both scientific knowledge and the specific technological know-how do not develop over time in the model, this denies the possibility of post-innovation improvements (Arcangeli, 1984) or learning processes (Antonelli, 1982). Moreover, these models are demand-pull models, since the potential demand is the critical variable in generating diffusion in each region, which gives no opportunities for supply to favour, or even limit, the diffusion process. A last deficiency concerns the fact that the application potential is known ex-ante, not allowing for the possibility of "*technological pluralism*", i.e. the possibility for a "dominated" technology to survive to become the "winning" technology in particular economic and spatial circumstances (Camagni, 1985).

The idea that innovation diffusion is a continuing process is the strong element of these analyses, and it has also been applied in theories of regional life cycle (Norton and Rees, 1979), which follow the well-known model of product life cycle developed by Hirsch (1967) and Vernon (1966). According to these theories, the regional differences in technological capabilities are the result of "physiological" processes due to the fact that technologies "grow old". This theory of technological development, also known as the "filtering down" theory, follows three stages (Vernon, 1966):

- the first phase relates to the initial production location of a new product and innovation. Because of the unstandardised final market and necessary input, the initial phase is characterised by a) flexibility concerning the input used; b) the difficulty of bringing the product onto the market. Thus, the initial phase is expected to take place in central and metropolitan areas;
- the second stage refers to the location of maturing products. The greater level of standardisation in the production process allows the strength of centripetal forces to decrease, and firms located in central areas may then locate part of their production in other advanced countries and regions;

- the third phase is related to the standardised product. Access to market information becomes less critical, thus allowing location to take place in developing countries.

Andersson and Johansson (1984) improved the explanatory value of the product life cycle theory by assuming different elasticities to distance of activities with various levels of knowledge (Table 3.2) (Kamann, 1986). In ranking levels of knowledge, they defined three levels, namely the information level, the knowledge level and the competence level. Creativity is then the combination of the two higher levels and is the ability to order and regroup information in an original manner. The spatial aspects of the various stages are presented in Table 3.2, being R&D, pilot production and standardised production each dependent on, respectively, competence, knowledge and information.

The positive consequence of the regional life cycle theory is that it is able to explain a very interesting phenomenon related to the spatial diffusion of new technologies, namely technological pluralism, by simply reflecting in a precise moment in time the inter-regional movement of technologies.

This theory, too, has some limitations. First of all, according to this approach, technological development is a straightforward process of transferring the same technology to a different location and exploiting the economic advantages of the area, but it completely ignores the subjectivity of the diffusion process, i.e. the interest, the ability and the receptiveness of a certain region in relation to a particular technological development. As Davelaar states:

> *"... the 'swarming' processes are 'creative' processes and not simple 'carbon-copy' processes of imitation" (Davelaar, 1991; p. 29).*

This idea was already present in the analysis of Freeman, who stated in his studies on innovation diffusion that this process cannot be viewed as one of simple replication and carbon-copy imitation, but frequently involves a string of further innovations, small and large, as an increasing number of firms get involved and begin to learn the new technology and strive to gain an edge over their competitors (Freeman et al., 1982). The limiting assumption is thus related to the analysis of a "constant" and given technology. On the contrary, technology changes over time, so that technological pluralism can be the result of the co-existence of different technologies at the same moment, and is not necessarily the result of the inter-regional diffusion of the same technology at different stages of its "life" (Figure 3.4).

Table 3.2

The product life cycle: spatial aspects of the various stages

STAGE	SUB-STAGE	RESULT	LOCATION	SKILL LEVEL	URBAN/ RURAL	MARKET
RESEARCH AND DEVELOPMENT	Fundamental Research	Discovery	University	High	Large Urban Areas	-
	Applied Research	Intervention	Firms' Labor.			-
	Development Prototype	Prototype	Firms' Labor.			test
PILOT PRODUCTION	Development and testing production processes	Production processes products	Near headquarters and R&D laboratories			introduction
STANDARDISED PRODUCTION	Automated, standardised mass production	Homogeneous product	Detached production plants	Low	Peripheral rural areas	Growth maturity saturation

Source: Kamann, 1986

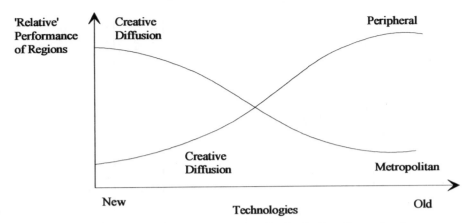

Figure 3.4. 'Creative diffusion' along technological trajectories in space

Source: Davelaar and Nijkamp, 1990

Another weak element of the life cycle analysis is related to the fact that despite the *intrinsic characteristics of the technologies under study*, the diffusion process always starts in metropolitan regions, which become the centre of all types of innovative activities. This assumption is justified by the fact that central regions have a higher attraction factor of innovative activities, but still the particular industrial practices and "*vocations industrielles*" of an area are widely overlooked. On the contrary these industrial practices could have a fundamental role in determining the receptiveness of a region to adopt a new technology.

Finally, an additional remark about this approach is the intrinsic assumption that the spatial structure of a region remains constant. In reality:

> "*in the long perspective, the economic structure of a region is not constant, but is being reshaped by new life cycles*" *(Davelaar, 1991; p. 92).*

As Metcalfe (1981) states, in criticising the static nature of epidemic diffusion models:

> "*a given innovation is diffused within an unchanging adoption environment, although there are well-documented reasons for expecting both innovations and environment to change as diffusion proceeds*" *(Metcalfe, 1981; p. 349).*

Development means "change" and not just a purely quantitative expansion of an existing economic structure, and therefore it happens only rarely that the long-term destinies of some areas are mechanically linked to their initial sectoral mix (Camagni, 1994).

The "creative" aspect, generally neglected in spatial theories on diffusion processes, is at the basis of the new "dynamic incubation theory" developed by Davelaar and Nijkamp (1990) for the explanation of spatial trajectories of innovation. This theory is a modern version of the traditional product life cycle model and overcomes some of its limits. The conceptual and theoretical framework of this theory is summarised in Figure 3.4, which relates the evolution of spatial structures to the effects of new technology systems and creative diffusion. According to this theory technology develops in time and space through three distinct phases (Davelaar and Nijkamp, 1990):

a) the incubation phase, related to the take-off of a new technology system; the effects in terms of "new" and "innovative firms" are expected to be first noticed in the larger and central metropolitan areas (availability of the skilled labour force and social overhead capital);
b) the catching-up phase. However, as the creative diffusion process proceeds along the "natural trajectories", on the supply side products become more and more standardised and the emphasis shifts from product to process innovations. Information flows and skilled labour become less important location-pull factors. Markets in metropolitan areas will first approach the saturation level. Standardised production will shift from metropolitan to non-metropolitan areas, while more advanced central areas will move on towards other products or production techniques;
c) the competition phase. When consumer markets become saturated and possibilities for further improvements of products decrease, price competition becomes more important. Non-metropolitan and peripheral areas find themselves in the best position as they receive already tested and fully developed technologies, and use them in a context of lower factor costs.

These innovation diffusion processes therefore generate the coexistence of different technologies in different regions, explaining once again the technological pluralism phenomenon. It can thus happen that while central areas are in a "stagnation phase", the periphery, on the contrary, experiences an ongoing intensive innovation phase, characterised by creative imitation (see Figure 3.4 above).

This interpretation, and in fact all the preceding spatial diffusion models, fits very well with the empirical evidence of the 1950s and 1960s, where technological change followed the historical diffusion of technological know-how through long waves of creative-imitation processes (Camagni, 1994). But more recently there have been signs of a more clear-cut turnover:

- firstly, at least in high-tech and advanced firms, the product life-cycle has drastically shortened, diminishing the time for a possible spatial diffusion, and its traditional curve presented by Abernathy and Utterbach (1978) lost its tail (Figure 3.5) (Camagni and Rabellotti, 1988). The process innovation curve is nowadays shifted dramatically upwards in the early experimental phases of the product life, and it is forced to follow the shape of the product innovation curve (Figure 3.6);

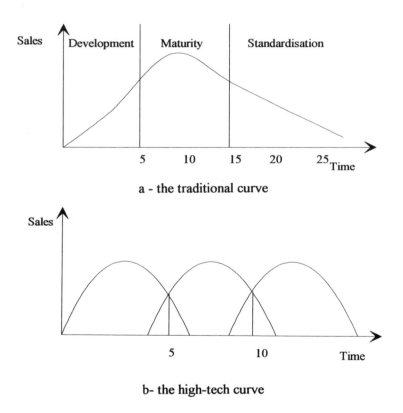

a - the traditional curve

b - the high-tech curve

Figure 3.5. The product life cycle

Source: Camagni and Rabellotti, 1988

- secondly, in more traditional industries (i.e. textiles, cars, watches, clothing, etc.) new technologies create the possibility for the "rejuvenation" of these sectors, through product improvements, giving central regions wide-ranging opportunities for economic restructuring;
- thirdly, fast innovation processes have proved to be possible through a closer interaction and synergy among the different functions of the firm, i.e. R&D, production, marketing, engineering, thus leading many firms to create special "mission units" devoted entirely to the innovation process. These "mission units" and special industrial agglomerations find their natural location in central areas (Camagni, 1988).

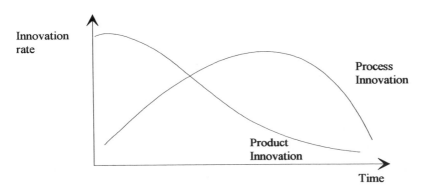

a - the Abernathy-Utterback Curves

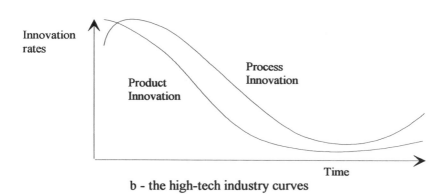

b - the high-tech industry curves

Figure 3.6. Innovation rates along the product life-cycle

Source: Camagni and Rabellotti, 1988

The nature of the present technological development thus seems to lead to a revitalisation and relocation of "traditional" production, from previously decentralised to core areas (Camagni, 1991c). These new characteristics of the present techno-economic paradigm seem to go against the conceptual framework of the product life cycle theory. The interpretation of regional disparities in innovative output is based on two distinct and complementary elements. The first one is labelled the "structural component" and refers to the fact that regions may differ concerning the extent to which their firms are innovative and linked to the field of technological change. The second component explains the innovativeness of regions. Identical firms have different degrees of innovativeness depending to their location in space. This latter effect is known as the production milieu component (Boyce, Nijkamp and Shefer, 1991; Camagni and Cappellin, 1984; Davelaar, 1991; Kamann, 1986; Molle, 1984).

This approach explains the relative innovative potential of regions on the basis of the regional endowment of firms having a high innovation potential (the structural component) and also on the basis of additional stimuli generated by an innovative environment (the production milieu component), this latter element being simply a revision of the concept of "selection environment" developed by Nelson and Winter (1977). While Nelson and Winter developed it by linking their analysis to the evolution of a specific technological trajectory, this concept has been enlarged to include the spatial element (Boeckhout and Molle, 1982; Molle, 1984). The wider definition of "selection environment" thus becomes "*that group of industrial and spatial elements operating in the sense of facilitating or limiting the evolution of a technological trajectory*" (Molle, 1984, p. 111).

These considerations lead to the remark that a micro-approach in regional development and regional disparity studies is in a certain sense necessary in order to understand regional innovative capabilities, through the firms' specificity in innovative processes as well as through the environment in which firms develop their innovative activities. In this respect, the approach developed by Camagni (1984) may be helpful. Developing an analysis at a micro-level, Camagni emphasised three critical elements for an innovation process to take place, namely the firm, the environment and the technology itself. The properties of, and the systematic interaction among, these three elements define the spatial innovative development. The ability of a *firm* to realise and exploit the potential of a technological and commercial idea, its receptivity towards information and its flexibility in a turbulent environment all depend on the *adoption of an adaptive internal organisation* as opposed to a mechanical, non-creative one (Burns and Stalker, 1961). In this respect, Molle (1984) underlines that the internal elements of a firm which facilitate innovations are:

a) *the firm dimension*, related to the internal capacity of a firm to develop R&D. In fact, the greater the dimension of the firm, the higher the financial possibility for R&D activity;
b) *the organisational philosophy of a firm*. To generate an innovative process it is of vital importance to have continuous contacts between R&D activity and the production and the marketing functions (Utterback, 1974), what Mignolet (1983) defined as the "receptivity of a firm to an innovation";
c) *the institutional structure of the firm*. Firms belonging to multinationals and very large firms, despite acting in advanced industries, can also operate on the basis of new products and processes "imported" from the headquarters. In this case, the local innovative degree depends on the degree of "freedom" and autonomy (Malecki, 1980);
d) *an already existing experience* in the use and exploitation of a similar technology. Firms tend to elaborate their innovative trajectories on the basis of past experience (Nelson and Winter, 1977).

Analyses of technological diffusion have shown that its speed depends largely on the internal properties of the innovation concerned. The following properties of the technology are the most important in this respect: its compatibility with the existing organisational structures; its complexity and appropriability; its advantage over the technology it replaces; its cost; and its communicability and pervasiveness in relation to potential adopters in different sectors.

The characteristics of the environment have been described as fundamental because they represent the economic and infrastructure preconditions necessary for the circulation of information. Moreover, the environment provides the basis for those psychological, cultural and social variables which define the level of education, the attitude to risk and to changes. Camagni (1984) underlines as preconditions for the adoption of innovation, in particular those situations when the following three elements are perfectly integrated:

a) the integration between technology and environment gives rise to the availability of information;
b) in the technology/firm integration there are many problems of compatibility with the already existing organisational structure and the relative advantage of the new technology over the old one. With a perfect integration between technology and firm the second precondition, namely the relative profitability of the new technology, is ensured;
c) the integration between the firm and the environment raises the problem of the cost of adjustment from the old to the new technology, a factor which is often underestimated in the analysis of innovation diffusion. For a new technology to be adopted it is not sufficient that it demonstrates superiority over the existing one. It is also necessary that the present value of the

differential earnings expected from the new technology is higher than the costs which have to be faced to bring the structure of the firm into line (Camagni and Cappellin, 1984).

This micro-approach analysis underlines once again how the study of spatial innovation diffusion processes cannot be pursued only on the basis of considering the information intensity in an area. Especially when the spatial dimension is included, there are other important preconditions for innovation adoption that have to be taken into account, namely profitability and adjustment costs from the old to the new technology. These latter preconditions may not be conducive to the rapid adoption of technological innovation in less favoured regions. In fact, the relatively low prices of production factors in these regions do not push towards labour intensive technologies, while the social, organisational and cultural obstacles to adoption raise the adjustment costs of highly profitable technologies (Camagni, 1991b). This is even more true when technologies under consideration are complex technologies, which have an impact on the organisational structure of firms. This is precisely the case with the telecommunications technologies, as we will see in the rest of the book. The adjustment costs from the old to the new technologies explain much of the spatial development trajectories of new telecommunications technologies.

Although this micro-approach is highly typical of all elements related to obstacles or incentives to innovation and even contains an explanation of the spatial dispersion of innovation trajectories, its limitation is related to its static nature. The majority of models on the spatial development of new technologies developed during the 1980s are in fact characterised by the staticity of the elements that are considered to be the causes of regional bias in innovative potentialities.

3.3. New technologies and regional performance

3.3.1. The role of technological changes in regional performance

The previous section has been devoted to the review of the literature concerning the spatial generation and diffusion of new technologies. The present section will examine the link between new technologies and regional performance.

The awakening of interest in the spatial development of new technologies during the 1980s is mainly related to the idea that technological development is becoming a fundamental explanation for regional disparities in economic growth. Thus, productivity, continuing innovation and technological advances are regarded as the driving forces for regional economic development

(Nijkamp, 1986a and 1987). An understanding of the laws by which technologies expand and grow through time and space is a must if the regions are to make a wise response to the different challenges they are facing, including the problems of take-off, sheltering local infant industry, and restructuring or reconversion (Blaas, 1991; Camagni and Rabellotti, 1990).

The logic in the assumption that technological change brings regional development in its wake is the simple statement that the adoption of new technologies allows for better economic performance via greater productivity of input factors. Within the framework of endogenous regional development, attention is oriented towards those elements that may change the rules of international or inter-regional competition, by altering the "entry barriers" to a market, as explained in traditional oligopolistic models (Modigliani, 1958). In fact, as is shown in Figure 3.7, the entrance into a market is impossible if the firm is facing an average cost curve such as AC_0 and a demand curve equal to D_0. The situation changes completely once the firm is facing an average cost curve like AC_1 and a demand curve D_1. These shifts can be generated either by consumers' income increases or by factors price decreases. Moreover, these curves' shifts can be generated by the adoption of product or process innovation (Camagni and Cappellin, 1984; Cappellin, 1983).

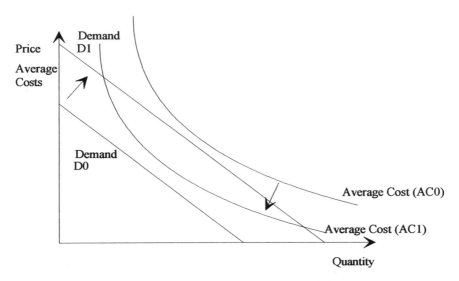

Figure 3.7. Evolution of entry barriers in an oligopolistic market

Source: Camagni and Cappellin, 1984

Process innovations are expected to act on the raw materials and labour costs, because of a rationalisation of their use, thus provoking an increase in the value added and a decrease in employment, i.e. an increase in productivity. Product innovation determines an increase in production in real terms or in its unitary price, thus acting again on the level of productivity. As Camagni and Cappellin state (1984):

> *"from the regional perspective innovation can be defined as every new resource allocation allowing for an increase in regional productivity"* *(Camagni and Cappellin, 1984; p. 148).*

In particular, technological change is regarded as the most probable and "modern" way to increase productivity, and consequently regional performance.

The importance of technology in the determination of regional performance explains why technology has to be developed even in less favoured regions (Camagni and Rabellotti, 1990; Gillespie et al., 1984). In an in-depth analysis of the reasons why advanced technologies should be developed in less favoured regions of Italy, Camagni and Rabellotti (1990) mention the following aspects:

- *the equity aspect*. New technology may show a (still unproved) potential for decreasing regional inequalities;
- *the efficiency aspect*. New technologies can generate a revitalisation of local industries and may increase their international competitiveness;
- *the increasing international competition aspect*. In a period of slow growth, international competition is destined to increase substantially, and thus the need for a strong national economy as a whole becomes crucial;
- *the continuing innovation aspect*. If new technologies are considered not only as process innovations, but more precisely as preconditions for continuing innovation processes, it becomes more evident why they should be developed even in less developed regions;
- *the product quality and market aspect*. On a micro-economic level, it is possible to say that even less developed regions cannot escape from introducing new technologies. If these regions present their products on the national or even international market, as they are expected to do because of the physical constraints of their local markets, these wider markets require sophisticated characteristics both in product (quality, design, novelty and variety) and in delivery conditions (time reliability, production flexibility), which can best be guaranteed by new technologies;
- *the comparative cost advantage aspect*. New technologies are undoubtedly labour-saving, and this could reduce the comparative advantage that less developed regions have with respect to labour costs.

In their study, Camagni and Rabellotti (1990) state that, although it is clear why technology has to be developed also in less favoured regions, it is less clear how it has to be developed. To reply to this crucial question of economic policy, one remark has to be stressed. In most theories technology appears as a neutral device, as a pool of opportunities available at a cost, as a certain kind of "public" good. But what really matters in the exploitation of new technologies and infrastructure, and what is overlooked in the literature, is the local cultural and organisational capability for exploiting these technologies and their potentialities, through a creative blending of technological devices, organisational styles and business ideas (Camagni, 1994).

The most advanced technology with the widest technological capability does not necessarily constitute the most appropriate technology for every region (Nijkamp and Priemus, 1992). Every region has to be considered as a diversified reality and the infrastructure endowment has to be tailored to each region's need. This statement is much more true once the analysis focuses directly on the possible diffusion of technological endowment to less favoured regions in order to decrease regional economic disparities. Even if it is clear that lagging regions cannot catch up rapidly in the production of new technologies, it is nevertheless fundamental that they must catch up rapidly in the utilisation of these technologies. As we will see in the next section, this is particularly true for advanced telecommunications technologies.

The "*intermediate technology model*" suggested by Schumacher (1965) as a response to problems of less favoured regions has been criticised for being a "pauperistic" and mainly "assistential" approach (Camagni and Rabellotti, 1990; Perez, 1985; Saraceno, 1981), and especially for not generating a long-term sustainable local economic development. On the contrary, the "blending" of the best technologies with industrial practices or "*vocations industrielles*" could create a sort of "*inter-regional pluralism of organisational models*", mixing advanced technologies with the specificities of local production traditions.

Under this perspective, three main groups of regions can be created, showing a different intensity of organisational and innovation capabilities:

a) *innovative regions*, typically central regions where proximity to information, to a skilled labour force, to high technological and scientific know-how and to potential demand facilitate and support innovation processes;
b) *adaptive regions*, where technologies are developed through the choice of an appropriate technology, which calls for an "appropriate design and adaptation" to the production needs;
c) *dependent regions*. In these regions local development relies on external technologies, available in the area through the location of branch plants of

multi-localised firms, thus generating a strong dependency on external decisions, skills and management practices that act in the long run against self-sustained local economic development.

During the last two decades a growing body of literature has studied the importance of physical infrastructure (i.e. transport and telecommunications) in the process of regional development and the reduction of regional disparities. From this perspective during over fifteen years of its existence, the European Regional Development Fund devoted 80% of its total resources to infrastructure. Nevertheless, regional disparities widened substantially in the 1970s, before stabilising again in the 1980s (Figure 3.8) (Camagni, 1991d; Vickerman, 1991).

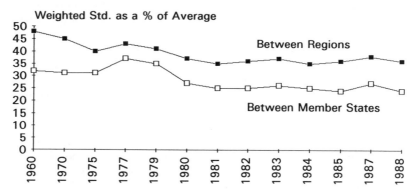

Figure 3.8. Evolution of income disparities in the EC, 1960-1988 (in ECU)

Source: Camagni, 1991d

It is not surprising that, for a long time, the role of infrastructure in the regional development process has been rather uncertain. As Vickerman (1991) underlines, it is obvious that dynamically growing regions have a well-developed infrastructure whereas lagging regions are typically deficient. However, long periods of regional development policies based on the creation of infrastructure in lagging regions have failed to make much impression on regional divergence. This result witnesses once again the fact that "infrastructure" itself does not inevitably lead to development and growth but it is a necessary rather than a sufficient condition for regional development. Other crucial economic elements have to be present in the region to support development (Button and Gillingwater, 1986a and 1986b; van Gent and Nijkamp, 1988; Rietveld, 1990; Rietveld and Nijkamp, 1992).

This might suggest that infrastructure is irrelevant to the real determinants of regional growth and development. This conclusion is quite wrong in our perception. The mistake is not in thinking that infrastructure leads to greater performance, but to assume a direct relation between the adoption of new technologies and a greater economic performance. In the same way as information does not necessarily mean adoption (see Section 3.2 before and Figure 3.4 above), so adoption does not necessarily mean greater economic performance. While, in the case of adoption, the crucial variable in the "cause/effect" chain has been defined as the spatial determinant, in the case of greater performance the crucial role is played by the spatial organisational and innovative capability (Figure 3.9).

The previous statement has far reaching implications for regional policy. For this reason and for the fact that it has been widely overlooked in regional analyses the present study will give special attention to it, especially regarding the new Information and Communication Technologies (ICTs) paradigm which has emerged since the last decade.

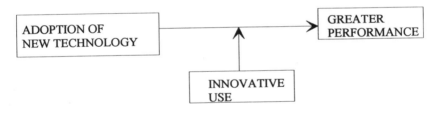

Figure 3.9. Linkage between adoption and greater performance

3.3.2. Network infrastructure and regional performance

New opportunities, but also new threats, have been created by the new Information and Communication Technologies (ICTs) paradigm to regional economies. ICTs are a catalyst for a set of structural economic transformations. As knowledge-based and information-based activities are becoming increasingly important "strategic forces", on which the competitiveness of firms and comparative advantage of regions will increasingly depend, a process of rapid spatial diffusion and use of these technologies is necessary and already well underway (Gillespie and Hepworth, 1986; Gillespie and al., 1989; Hepworth and Waterson, 1988).

The new information capital embedded in the modern and fast computer networks opens many possibilities for innovation, namely network-based innovations, fostering indigenous regional development. The study of the effects on regional performance via the exploitation of new information networks can be related to the general theories on the role of infrastructure on regional development. As Bruinsma et al. (1989) and Blaas (1991) report, three kinds of theories can be distinguished in this field:

a) the first are the general development theories, including both the export base theory and the classical growth theories. The former is based on the assumption that a region's growth is dependent on the region's basic production activities (i.e. the production for export out of the region to other regional or international markets). A well-developed infrastructure is one of the factors supporting the export ability. The latter theory is based on a macroeconomic production function. The economic growth of a region is seen as being dependent on the supply of capital and labour. Differences in regional economic growth are explained by the inter-regional mobility of these production factors and mobility itself is dependent on the quantity and quality of the available infrastructure (Clark, 1940);

b) the second kind of theories dealing with regional growth and infrastructure develop the idea that infrastructure is the regional engine for growth, one of the most strategic production factors. The regional development potential theory is an example in this respect. Already developed by many scholars (Biehl, 1986; Blum, 1982; Bruinsma, Nijkamp and Rietveld, 1989; Nijkamp and Priemus, 1992; Vickerman, 1991), this theory claims that regional development potential is determined by an interplay of mobile production factors and potential factors. Both the quality and the quantity of infrastructure are factors which determine the attractiveness of a region. Table 3.3 reports a selection of studies of this kind (Bruinsma, Nijkamp and Rietveld, 1989; Rietveld, 1989; Van Gent and Nijkamp, 1988);

c) the third group of theories refer to the development theories based on transport systems. These theories put a particular emphasis on the role of transport infrastructure in the explanation of regional development and performance. Improvements in transport systems lead to (hardly reversible) spatially differentiated effects, favouring regions receiving a greater infrastructure endowment.

The study of the effects of the new information and communications technologies can be pursued in each of these theories. In the first group, just as transport infrastructure supports basic production activities so do communication networks, by facilitating production through an increase of information flows into the region which allow new market opportunities to develop. In the second group of theories, the role of the new communication infrastructure is very simple to determine: many studies have focused their attention on the role of new computer networks as an attractiveness factor for business related activities. Even the third group of theories can also be simply related to the telecommunications infrastructure: accessibility to information and possible reduction of costs in information collection are seen as the major location factor for this techno-economic paradigm.

Table 3.3
Examples of the production function approach to infrastructure modelling

Author	Country	Number of sectors	Number of types of infrastructure	Presence of: Labour	Private capital	Form of Production function
Biehl	E.C.	1	1	yes	no	Cobb-Douglas
Blum	F.R.G.	3	8	no	no	Cobb-Douglas
Andersson et al.	Sweden	1	7	yes	yes	Cobb-Douglas
Snickars and Granholm	Sweden	21	5	yes	yes	Leontief
Nijkamp	NL	1	3	yes	no	Cobb-Douglas
Fukuchi	Japan	3	3	yes	yes	Cobb-Douglas
Kawashima	Japan	8	1	yes	no	linear

Source: Bruinsma et al., 1989

Despite the strong similarities between transport systems and telecommunications networks in the role they play in regional development, the case of telecommunications infrastructure deserves particular attention because the advantages computer networks can generate on regional performance go beyond the well-known positive effects expected from an infrastructure development. Theoretically, new computer networks are a vehicle for two kinds of "economic advantage":

- the first, in common with the other kinds of infrastructure, is related to the fact that infrastructure may generate various economic effects in terms of value added, productivity and employment, through direct (conventional Keynesian expenditure effects) and indirect effects (via intermediate deliveries) (Bruinsma et al., 1989; Rietveld, 1989; Van Gent and Nijkamp, 1988);
- the second advantage stems from their nature of "networks". They physically link different regions (or actors) together, which benefit from those advantages generated by synergies among regions (or actors) operating in different economic environments, by the availability of more information flows (thereby reducing uncertainty accompanying decision-making processes in a world of imperfect information), by exploiting possible network-based innovation, and by achieving previously unknown markets. It is possible to achieve all these advantages by the existence of a physical linkage on the network, i.e. by what we called before (Chapter 2) the positive effects of "network externalities". These advantages increase with the number of subscribers entering the network.

The general idea underpinning these statements is that the connection of a new subscriber generates advantages for the already existing users of the network. These advantages, from the economic point of view, are an increase in information availability, in synergies with other actors, in gaining previously unknown markets, in obtaining complementary production inputs, and in exploiting network-based innovations.

Being networked matters, and it generates advantages and positive externalities even from the economic point of view (Camagni, 1991b and 1992a). However, this statement has a limit, underlined by the fact that even a tool such as a telecommunications network generating positive externalities can turn out - in some spatial circumstances - to be an instrument for new forms of spatial exploitation (Gillespie, 1991). If it is widely recognised that less favoured regions suffer from not being networked, they can also be disadvantaged by the opposite situation, viz. from being in a network. In fact, less developed regions can be excluded from leading the network and thereby from participating in the definition of the "rules of the game", because of their low capacity for exploiting network externalities. Instead, they could play a role of servants in the network, with direct negative effects such as the loss of control and power on local markets, on local resources, or on locally-based information.

It has become a widespread belief in the literature (see Section 3.2) that technological innovation is not a "deus ex machina", but can be induced by policy intervention. This awareness has generated a crucial question for policy makers: Which type of policy is needed for which type of technology in which area? The right reply to this question presents the key of success for technology-oriented regional policy.

The present work will try both to create a more solid theory on the effects of network externalities on regional growth and disparities, and to measure these effects empirically. Such an attempt has so far never been made.

3.3.3. The role of ICTs in regional disparities

The adoption and use of new advanced communications technologies have very deep spatial implications which can be analysed on two distinct levels (Hepworth, 1987 and 1989; Kellerman, 1993):

- a macro-level, by understanding the effects ICTs have on regional development;
- a micro-level, by measuring the potentialities of these technologies for reshaping regional restructuring through decentralisation or centralisation processes of industrial activities.

At a macro-level the issue is thus related to the understanding of ICTs' capacities in fostering indigenous regional development. Two different conflicting theories emerge:

a) the first theory relates to the idea that these communications innovations can be used to overcome some of the previous problems of peripherality, such as limited information environments, remoteness from markets, or lack of access to specialist services. In a sense, the advantages which are perceived to accrue to "centres" in national space economies, i.e. "agglomeration economies" in one sense or another, have the potential at least to be increasingly spatially diffused through ICTs;
b) the second, and opposite, theory is based on the idea that the diffusion of new technologies is a centripetal process, destined to occur *a priori* in central areas, where a high potential demand density is present. Hence, backward areas will be penalised in the spatial distribution of these new technologies for a series of reasons:

  - lower level of potential users,
  - lower effectiveness of "network externalities" in generating cumulative adoption processes. The interest of a new potential user is in fact higher if he can be linked to central areas, rather than to the periphery;
  - lower possibility to achieve a critical mass of adopters, a crucial key variable to generate a rapid cumulative adoption process.

A limited access to networks and services thus constitutes the major threat to peripheral regions. From this perspective, the shift towards demand-led criteria governing investment decision-making with respect to network modernisation and the introduction of new services will penalise the periphery. The result of such an institutional change will in fact begin to exert a much greater influence on the geography of service supply, which will inevitably become more geographically differentiated. Moreover, the implicit cross-subsidies involved in pursuing the universal service principle, whereby uniform changes are levied regardless of the actual cost of providing the service, come under pressure in more liberalised supply regimes (Gillespie et al., 1989). The consequent outcome is a movement towards a "cost-causative" price system, in which charges differences reflect the real costs of providing a given service to a particular location.

Undoubtedly, the limited access to networks and services constitutes the major threat to peripheral areas, which could be excluded from enjoying the advantages associated with these technologies. Nevertheless, once they obtain an infrastructure endowment, major opportunities are given only by innovative use, through the choice of an "appropriate technology". By this term we refer to the more advanced conception of "appropriate technology", calling for an "appropriate design and adaptation" of the best technologies to the production needs and "*vocations industrielles*" of the particular area (Camagni and Rabellotti, 1990). The blend of the best technology elements with traditional industry practices ensures a more sustained local development, in contrast with the "intermediate" technology model suggested by Schumacher (1965). Following these innovative adoption patterns, peripheral regions become "adaptive regions", rather than "dependent regions", where the difference between the two categories lies in the way technologies are used and exploited (see Section 3.3.2).

As we have already mentioned, from the point of view of lagging regions, even if it is clear that they cannot catch up rapidly in the production of new technologies, it is nevertheless evident that they can - and also must - catch up rapidly in the utilisation of these technologies. They sell their products in an international market, and this market requires sophisticated characteristics both in products (quality, design, novelty and variety) and delivery conditions (time reliability, production elasticity for peak or unexpected demand) that is hardly possible to achieve through traditional production technologies (Camagni and Rabellotti, 1990; Goddard et al., 1987). A strategy of a blending of the best technologies with more traditional and organisational practices seems the most effective here (Camagni, 1991c).

At a micro-level, ICTs also have the potential to reshape the spatial structure of industrial activities (Capello, 1993). Two opposite ideas govern this field:

a) according to the first view, ICTs provoke strong decentralisation effects, by allowing on-line remote production control, on-line logistical systems and on-line administrative management systems;
b) according to the second and opposite view, the spatial restructuring of economic activities follows a centripetal pattern, favouring more advanced firms and regions, this process being justified by external economies, by spin-off effects of the skilled labour force and by technological and organisational know-how required to develop an innovative use of these technologies. The highly modern infrastructure endowment of central areas acts as a strong location factor for firms.

These two contrasting principles can be analysed through empirical evidence on the present behavioural trends in some firms. These technologies can be considered "enabling technologies", representing necessary but not sufficient conditions to generate spatial restructuring and thus to diminish the "tyranny of territory" (Gillespie and Hepworth, 1986; Nijkamp and Salomon, 1989). Vital for the development of a new spatial production restructuring is the exploitation of these technologies alongside organisational decisions and location choices of the firm (Camagni and Capello, 1993).

The results of the empirical analysis underline two aspects (Capello, 1993):

a) first of all, new technologies do not result in an immediate spatial reorganisation of production; they operate in the sense of consolidating the already existing spatial structure;
b) secondly, both organisational and strategic decisions have to be taken into account in the process of spatial relocation.

In the long run some preconditions for spatial reorganisation will be developed which even today already exist in the sense of a constant location of productive activities in space, namely:

a) the modern and advanced telecommunications infrastructures will achieve a capillary geographical extension over the national territory, which will operate in the sense of physical unification of all possible geographical locations;
b) the organisational structures of the firms will be more flexible ensuring the integration between organisation and technology, which is strategic for the innovative use of these technologies;
c) technologies themselves will be more "integrated", i.e. software and hardware technologies will be integrated within functions and with the organisational aspects of the firm. In this way the new forms of organisation will be developed. As a result of these preconditions, a new spatial "industrial village" may emerge. The "just-in-time" organisational

form will still require a physical proximity between suppliers and producers, and from producers to the final market. In this case, we can expect to achieve a spatial clustering of economic activities as with traditional location theory (Swyngedouw, 1987).

Only with the development of these preconditions it is possible to think about a new spatial restructuring. Profound changes will happen only in the long run, when the present locations, still highly dependent on traditional and historical location patterns, will be influenced by new corporate strategies, by new organisational choices and by some specific preconditions. This new spatial configuration will undoubtedly take into account the existence of new technologies capable of destroying physical distance between different industrial activities.

3.4. Conclusions

This chapter surveyed the vast literature existing on the spatial aspects of technological change. This literature may be classified under two different categories, according to the subject of the analysis. Many studies have in fact dealt with the *spatial adoption* of new technologies, where the primary object was to analyse the territorial trajectories of a new technology. This field of analysis started with the well-known Hägerstrand model, where the technological trajectories over space follow the spatial dispersion of information. The limits of this pioneering contribution have been emphasised in many recent studies which particularly considered the spatial capacities of absorbing and accepting new technologies. These capacities stem inevitably from the spatial determinants of an area, i.e. its economic and social characteristics.

The contributions to this field of research have greatly varied, concentrating on either the region and its capacity of absorbing technological changes or the technology itself and its spatial pattern. All come to a common conclusion that innovation processes follow a centripetal process, starting in the centre, and then they subsequently develop over space, when information and know-how spread around.

A second field of analysis in the area of the spatial aspects of technological change is related to the impact that these technologies have on *regional performance*. The general idea developed by these theories is that innovation improves regional development and that for this reason it may even become a strategic economic policy tool for reducing regional disparities. In this respect, our idea is that the mere adoption of advanced technologies does not necessarily lead to better regional performance. Regions have to achieve an innovative use of these technologies in order to exploit all the economic advantages they provide. In this study, we will prove this idea at an empirical level in the case of telecommunications technologies. In other words, what we argue is that some seedbed conditions are necessary in order to achieve better regional performance through technological changes.

Our analysis on network externalities will be developed at a regional level and we will test empirically (Chapters 7, 8 and 9) whether network externalities develop under certain seedbed conditions. From the review of the literature presented in this chapter we would expect that:

a) telecommunications technologies are more easily developed in advanced regions, responding to the centripetal forces existing at the spatial level;
b) network externalities are not exploited in the same way in advanced and backward regions. The main literature on the spatial aspects of technological change leads us to expect that network externalities are likely to be more exploited in advanced regions;
c) finally, we would expect these technologies to have a positive impact on the regional performance only under certain preconditions. If these preconditions are not present, network externalities cannot provide positive effects on the spatial performance.

In this study we will be able to test these hypotheses and come to some concluding remarks which assume a particular interest at the normative level (Chapter 10).

Notes

1. For a systematic presentation of these works, see Davelaar, 1991
2. See Capello (1988) for the statistical results.

4 Industrial and spatial performance in the presence of network externalities

4.1. Introduction

The interest in the analysis of network externalities and in their effects and consequences stems from the role that telecommunications have in economic systems, influencing both the static efficiency of economic systems as well as their dynamics in terms of competitiveness of both firms and territorial systems. In fact, it has been widely recognised that the allocation of given resources to given economic functions, and the generation and introduction of technological and organisational innovations increasingly depend on access to advanced telecommunications networks (see Chapter 3).

Especially in a dynamic context, the value of advanced telecommunications networks and services to firms in both the industrial and service sectors is a consequence of the key role now played by information. Historically, technological change has affected the "production component" of the organisation, i.e. the handling and processing of material resources, and the competitive advantage of firms has changed from a labour-intensive production process to one which is capital-intensive. Nowadays, however, technological change is affecting the "organisational component", i.e. the handling and processing of information, and competitive advantage is increasingly being underpinned by the development of information-intensive activities. From this perspective, it is of fundamental importance to study telecommunications networks development, and especially the economic principles governing their development.

In Chapter 2 we indicated that there is an increasing awareness of the fact that network externalities are the most significant mechanisms explaining the diffusion of interrelated technologies, such as telecommunications networks and network-based services. The basic idea - present in the literature - is that once the critical mass has been reached, the diffusion process proceeds to develop with ever higher diffusion rates among potential users. However, only a few attempts have been made to measure these mechanisms, while in particular no effort has been made to see whether network externalities are exploited by particular firms and are induced by particular seedbed areas. In fact, it may very well be that network externalities influence the industrial and spatial diffusion processes, reinforcing or hindering the traditional technological and spatial dynamics.

The effects that network externalities generate on the performance of firms have so far been neglected and this chapter provides a conceptual analysis of the effects that the existence of network externalities produce on the industrial performance of firms which use these technologies. In the present chapter a conceptual framework is developed and used to analyse the expected industrial and spatial performance due to advanced telecommu-nications technologies. The aim of the framework is to make a first step towards the integration of the spatial aspects of technological changes and network externalities theories, these latter being considered as *the* main driving forces for the development of modern ICTs technologies. However, it should be stressed that we do not intend to ignore the role of other "driving forces" identified in the literature on spatial economic development. Our chief purpose, however, is to underline the role of network externalities as the main features characterising the dynamics of industrial and spatial performance.

This chapter contains an analysis of possible reasons (i.e. micro stimuli) for the diffusion of telecommunications technologies among firms (Section 4.2), in order to relate our study to the more general conceptual approach known in the literature. In Sections 4.3 and 4.4 our own conceptual framework on the linkage between network externalities and the performance of firms and regions is presented. Section 4.5 contains an analysis of costs and benefits of network externalities and provides a typology of their effects.

4.2. Micro-stimuli for telecommunications networks adoption

Our framework is based on many studies which have been grounded on the premise that technological change and innovation are of vital importance for the improvement of firms' performance, via an increase in productivity (Stoneman, 1983, 1986 and 1992). All literature dealing with innovation diffusion processes recognises the linkage between innovation and better performance, although it is still fair to state that the majority of the empirical work on this topic has not yet caught up with the theoretical advances.

Innovation and technological changes are thus more and more considered as supplying the impetus for economic development. The abundant literature dealing with the diffusion of technologies among firms, and especially explaining and analysing the most important reasons for the diffusion of innovative technologies, may be synthesised as follows according to the main reason for the adoption processes (Stoneman, 1992):

a) *availability of information*. The diffusion of new technologies is explained in terms of information diffusion among potential users. Late adoption of new technologies is justified and explained by the late penetration of the necessary information. This interpretation of innovation diffusion is studied in the so-called *epidemic models*, which are rather widespread in spatial analyses (see also Chapter 3);

b) *differences in corporate structure*. The different diffusion of a technology among firms depends on the fact that firms are different, with different capacities for financing, using and exploiting these technologies. This idea suggests that the larger the firms, the quicker the adoption, because of their greater capacities to exploit economies of scale. Models explaining the diffusion process of innovation on the basis of the structural characteristics of firms have been labelled *rank models*;

c) *profitability of the new technology*. It is of course very simple to understand that the higher the profitability, the higher the adoption rate. However, as the use of a technology increases, the cost of production of this technology decreases, output expands and price falls. These changes will affect the return from the adoption of new technology. At a given cost of acquisition, there will be a number of adopters beyond which adoption is not profitable. Models exist, which explain the diffusion process of innovation among firms on the basis of the profitability of the technology itself. These models are known as *stock models*;

d) *the number of adopters*. The first adopters are the ones who obtain the greatest profitability from the adoption because of the exclusivity in using these technologies. Profitability is a decreasing function of the number of adopters. Some models use the number of adopters as the main explanation of diffusion processes of innovation and are called *order models*.

All these (epidemic, rank, stock and order) models explaining the reasons for adoption each represent a possible interpretation of the real world. They do not contradict each other, but simply underline a different aspect of the same phenomenon. Their empirical validity in general, has still to be tested[1]; however, they do represent a logical explanation of the innovation diffusion processes. In the case of telecommunications technologies one qualification needs to be made. Only the epidemic models and the rank models are entirely valid in the case of telecommunications technologies. In fact, one may easily argue that the greater the information on the way these technologies work and can be used, the greater the adoption rate. Moreover, the greater the capacity of a firm to finance and exploit these technologies, the greater the adoption rate. However, the same does not hold for the stock and order models. In fact, in the case of telecommunications technologies, the value of the technology itself increases with the number of adopters. For the same reason, it is not at all true that the first adopters will be the ones obtaining greater productivity from the technology. In the case of telecommunications it is rather a reverse link, that *the greater the number of adopters, the greater the profitability from the technology*.

Bearing this consideration in mind, our framework of analysis presented in the next pages can be linked to the "order model" (Karshenas and Stoneman, 1990). In fact, in our model the basic explanation for telecommunications network development among users is the number of adopters. The micro-stimulus for adoption is the number of already existing subscribers of a network. However, as we said before, in the case of interrelated technologies, the higher the number of already existing subscribers, the higher the profitability a new adopter achieves.

With our framework, we do not intend to deny the importance of other factors which may act as the driving forces of development. We simply want to underline, and test empirically in a subsequent chapter (see Chapter 7), the role network externalities have on the development of telecommunications technologies.

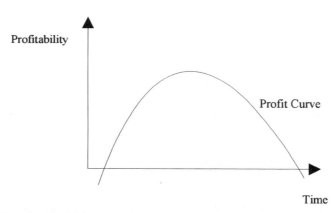

Figure 4.1. Profitability curve

This means that in the context of a graphical representation of the profitability curve in time (Figure 4.1), its shape would always be the well-known profit curve, with an increasing slope (at decreasing returns), followed after a certain time by a decreasing slope. The difference with the traditional model is that the reasons for the cumulative starting phase are not the common learning processes and very low transaction costs, but primarily the existence of a high number of users. This means that the main reason explaining the micro-stimuli for the adoption of these technologies is the achievement of non-paid for advantages in the network, since the profitability of interrelated technologies increases with the number of users. In fact, the advantages adopters have in the use of these technologies are strongly dependent on the number of users.

The micro-stimulus to innovate is thus explained for interrelated technologies by a high number of already existing users. Network externalities are then interpreted as the motive for joining a network (Figure 4.2).

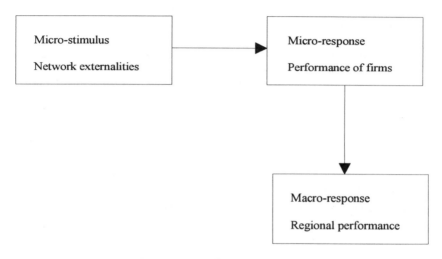

Figure 4.2. Conceptual framework of network externality effects

4.3. The economic symbiosis concept

4.3.1. The framework

Our basic research questions are related to the linkage between network externalities and industrial and regional performance. In particular, our main aim is the analysis of the question whether or not network externalities may be measured in terms of industrial (i.e. micro-) and regional (i.e. macro-) performance. Such a research question is fraught with many empirical difficulties of a methodological nature, which will be dealt with in the empirical chapters of this study. At the purely conceptual level, however, it is easier to envisage a positive relationship between network externalities and the corporate productivity.

The achievement of greater economic performance by exploiting benefits derived from joining a network is what we call an *"economic symbiosis"* effect, an improvement in the economic performance based on non-paid for synergies among firms. The "economic symbiosis" may be better defined on the basis of a set of firms strongly interrelated to each other via a physical network. This set of firms and its interdependent sectors have a relatively high productivity because of the achievement of strong advantages in comparison with the non-networked firms. These advantages may be classified in terms of direct and indirect advantages (Table 4.1). The definition of direct advantages is related to the fact that the advantages a firm gets via a network directly affect (positively) the productivity of a firm. These can be classified as:

- *static advantages*, which may be summarised in synergies among actors operating in different economic environments;
- *dynamic advantages*, represented by the possible achievement of network-based innovation, and of previously unknown markets;

Other advantages may be achieved via a network, which indirectly affect the productivity of firms:

- *static advantages*, such as information provision induced via network interconnection, and;
- *dynamic advantages*, such as complementary assets.

Table 4.1. Typology of advantages of network externalities

	STATIC	DYNAMIC
DIRECT	Synergies among firms	Achievement of network-based innovations and of previously unknown markets
INDIRECT	Information provision	Complementary assets

These advantages are generated by the existence of a physical linkage on the network, i.e. by what we call "positive network externalities". These advantages are expected to generate a positive effect on the performance of firms. As a result of more and better information, more synergies with other sectors operating in different economic environments, a higher degree of innovativeness in bureaucratic procedures, and more complementary assets, these industries are more efficient (in terms of productivity) than others. The synergetic and "symbiosis" effect generates a set of advantages the firm receives from being networked but does not pay a marginal price for them. Thus, via a network, both pecuniary and technical externalities may be achieved and exploited. *Ceteris paribus*, this set of networked firms is capable of generating dynamic growth in the economy, via large spillover and multiplier effects.

As mentioned above (see Chapter 2), the "economic symbiosis" model may show many similarities with the "growth pole" theory of Perroux (1955). In Perroux's approach, a set of firms, strongly interrelated to each other via input-output linkages around a leading industry (*L'industrie motrice*), is able to generate strong cumulative and multiplier effects on the economy via spillover effects. The effects of a "growth pole" and of an "economic symbiosis" are very similar, indeed. The positive and cumulative effects of a set of industries are the outcome, however, of two different phenomena. In the growth pole theory, the determinants of the strong dynamic effects are the existence of advanced technological practices and high innovation rates in this set of industries. In the "economic symbiosis" approach, the dynamics of these industries is explained by the physical linkage among firms, generating advantages such as more synergies among economic operators and more information. The access to a network and the non-paid for advantages a firm gets by joining the network play a crucial role in the performance of firms, primarily via an increase in productivity. This assumption is evident, since a firm may obtain advantages from being networked, without paying a marginal price for them. In other words, the (technical and pecuniary) network externalities generated via the network are the intrinsic explanation of the "economic symbiosis" phenomena.

Although there may be similarities in effects reflecting a better performance of firms, the causes or main reasons for this phenomenon may have completely different origins and a completely different nature. Without denying the existence of other major factors which improve the performance of firms (i.e. international market developments, better marketing strategies, etc.), the present framework aims at focusing only on one distinct cause of increased performance, viz. production network externalities. At the conceptual level, the framework thus rests on the assumption that the main micro-stimulus explaining adoption processes of advanced telecommunications technologies is the existence of network externalities (see Figure 4.2 above).

An interesting question at this time is whether the "economic symbiosis" effects of a physical linkage are positive for every firm joining the network. The existing literature on this issue (see Antonelli, 1990; Camagni and De Blasio, 1993) tends to attribute a positive effect to each new entrant in the network. The issue is, however, rather complex and needs further investigation and discussion, which will be presented in the next sub-section.

4.3.2. Micro-conditions for the exploitation of "economic symbiosis"

The (micro-) conditions for the exploitation of the "economic symbiosis", analysed in greater detail in this section, depend on different elements, namely:

a) the nature of networks (public, private, club networks);
b) the role of firms in the network.

In this section we will analyse these two aspects in detail.

The nature of networks. A crucial role in defining the effects network externalities have on the performance of firms is played by the nature of the network. As mentioned before, in the literature there is a tendency to attribute positive effects to all subscribers joining a network, with no regard for the actual nature of the firm under consideration.

In Chapter 2 we defined different groups of networks, i.e. private, public and exclusive networks. The intrinsic characteristics of public and shared goods of the last two groups contain in themselves the features for the existence of network externalities. This is not the case for private networks. In fact, their nature as private goods prevents free access to new subscribers. These networks are meant to be used by a single firm (among its different departments) with no access for other external actors. In this specific case, network externality effects cannot be achieved, because the advantages a firm gets from being networked are reflected in its marginal price (see Figure 4.3) and common advantages of joining the network are not meant to be there.

The effects on the performance of the networked firm may be positive, since the firm may achieve a better internal efficiency in its bureaucratic procedures. Nevertheless, its performance increase has nothing to do with the "economic symbiosis" effect. This remark is important when developing an empirical analysis, since a study of the network externality effects of a private network on the firm's performance would be rather misleading in its results. Better performance in that case could not be explained by positive network externality effects, but mainly by higher internal efficiency and productivity.

A strong difference also exists between public and exclusive networks in the possibilities for users to exploit the advantages stemming from network externalities. For exclusive networks, members joining the network obtain advantages in comparison with non-users. As the theory says, the voluntary linkage with an exclusive network is possible only if the entry costs are less than the benefits deriving from joining the network. The immediate consequence is that network externality effects are generated when a member joins a network. In other words, new members receive advantages straightaway from their participation in the network, despite their role in it and, moreover, they themselves contribute to these advantages by the very act of becoming members.

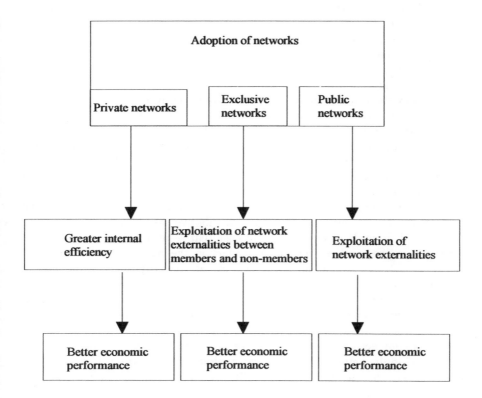

Figure 4.3. Relationship between network adoption and corporate performance

This conclusion cannot be drawn for public networks. The access to the network is free for everybody, and thus the access to it is not the discriminating factor for the exploitation of network externalities. The discriminating feature in the exploitation of network externalities, the quasi-private good which allows subscribers to use and exploit network externalities more than their competitors, is the way in which these technologies are used, viz. the organisational capability of integrating their potentialities with the needs of firms and regions. In fact, the simple adoption of these technologies does not necessarily mean the full exploitation of network externalities.

This remark leads to the second element of our analysis, i.e. the role a firm plays in the network.

Role of a firm in a network. As we have already mentioned, the literature on networks tends to be rather optimistic in saying that if a firm enters a network, it will surely benefit from it. This implies that networks always generate positive network externalities (see also Section 4.5).

This is rather a misleading approach and would be true only if firms were homogeneous. On the contrary, the capabilities of exploiting network externalities are far from being equal among firms and are directly dependent on the way these technologies are used. Firms can in fact achieve a status of co-operation via a network, which means the exploitation of *mutual reciprocal network externalities*. Nevertheless, it is also possible that a different kind of relationship is established, more dependent on a hierarchical structure among firms, which could generate a sort of dependency of one firm on another, and then a certain kind of "exploitation" takes place. In this case a *unilateral network externality* effect is generated.

This is true when we deal with public networks, where everybody has legal access to the network. In this case distributive issues arise even among networked firms. It can easily be the case that some firms, because of their strength in the market, behave as leaders in the network and manage to achieve a better economic position through the exploitation of weaker partners. With exclusive networks, the same distributive issues of economic advantages among users arise, when only members of the network are taken into account. Even in this case, the analysis can be the determination of:

- who, among network members, has the discriminating features for acting as a "leader" in the network, and thus can exploit network externalities more. An example of this situation could be a firm acting as a leader in the market. Its leading business position would ensure that the firm chooses the network standard and imposes it on its suppliers, customers and competitors. The choice of the standard may stem from the existence of a technological and organisational know-how in the use of that particular standard. This know-how may put the firm in the position to be the leader of the network;
- in the case of competition among different exclusive networks, whether it is more worthwhile to keep the distinction among exclusive networks, acting as competitors, or whether members gain more from removing barriers to entry among competing networks by creating a wider network.

In this latter case, the exploitation of network externalities depends on firms' expectations about the degree of "co-operation" or competition of other firms in the network. If they expect that other firms will be willing to cooperate, then the degree of physical connectivity will increase, and, consequently, the benefits that firms receive from their connectivity will increase, too. Examples are, for instance, telex and telefax networks. Apart from going through a normal product life cycle, their evolution also depends on the intensity of use by other clients (i.e., the potential access a network offers to designated parties of a firm) (see also Rietveld et al., 1993). This also raises the issue of compatibility of the technology of the network. If networks cannot be made compatible, they are by definition competitors and face fierce competition in connection with their introduction and market penetration. An important issue is to identify the degree of mutual openness of two "exclusive networks". The critical point is to define whether it is more convenient for an exclusive network to open the access to the members of other exclusive networks. The advantage of achieving a new information source is reciprocal in the case of a compatibility of two formerly closed networks and thus the decision depends on the willingness to lose unique access to strategic information compared with the willingness to gain new information.

An illustrative example is presented by the case of the flower auction network in the Netherlands, which provides all strategic information on the export market etc. to all flower suppliers who are members. A great problem is now presented by the possibility of linking the national Dutch network to flower database networks in other countries. The competitive behaviour of monopolising information is in contrast with the co-operative behaviour of joining a larger network with more competitors, which at the same time also guarantees better international market access.

The final decision on free access to a network depends on two kinds of expectations:

a) expectations on the role that firms already belonging to the network will play in the market. The stronger the firms are, the higher the degree of openness. In this case the already existing firms are expected to play essentially the role of winner in a wider network;
b) expectations of the already existing firms in the network concerning the behaviour of firms which do not yet belong to the network. In this case, if expectations are linked to competitive rather than co-operative behaviour, the possibility of opening up the network is rather low.

It is in fact not always true that the higher the number of subscribers, the higher the benefits a firm can get[2]. The explanations are twofold. The first reason is that power relationships have still to be defined between firms with new relationships, and the uncertainty about the potential economic power relationship rather reduces the ability to appreciate the advantages one can get by being linked to a network. The second reason is that network externalities will already exist if a business relation is established before the physical linkage among the two firms. As in the case of public networks, externalities are in fact generated by connectivity, but this connectivity is the result of an economic interest rather than purely a physical communication possibility.

4.4. The spatial symbiosis concept

4.4.1. The framework

The existence and exploitation of network externalities are not confined to the firm level. The "economic symbiosis" concept can in fact be translated into a spatial setting. This can be done, first, by the simple extension that the set of networked industries is spatially clustered, and, secondly, by focusing on spillover effects not for the economy as a whole, but in the immediately surrounding hinterland.

In other words, a "*spatial symbiosis*" effect takes place when a set of "networked firms" are present in a specific region (Figure 4.4). This leads to a non-zero sum of the positive effects on the firms' performance generated by the non-paid for advantages of being networked. A set of spillover effects is in this way generated on the local (regional) economy, which can be measured in terms of better regional performance. This phenomenon can give rise to cumulative effects *à la* Myrdal (1959) and can guarantee a local sustainable development.

The spatial symbiosis concept is very similar to the "growth centre" theory of Boudeville (1968), who first provided a "translation" of Perroux' growth pole theory in geographical rather than economic space. Boudeville's theory emphasises the importance of agglomeration and concentration of economic activities for local economic growth. The geographical clustering of a set of innovative industries produces strong backwash and spillover effects over space, which may lead, in the presence of some local prerequisites (i.e. a highly developed infrastructure, provision of centrally supplied public and social services, a demand for labour) to local economic growth.

The difference between the "growth centre" and the "spatial symbiosis" concept emerges quite clearly. Again, it is reasonable to claim that, although the local economic effects are the same, the nature of these phenomena are quite different. In Boudeville's theory, local economic growth is explained essentially by the geographical agglomeration of innovative firms. In the "spatial symbiosis" approach, it is not only the presence of a set of networked firms which is the reason for the generation of local economic growth.

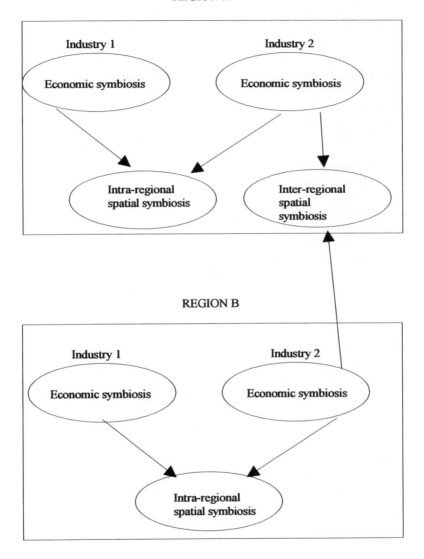

Figure 4.4. Economic and spatial symbiosis

Within this approach, a regional economy may also very well benefit from the input obtained by inter-regional networked firms. In fact, it is likely that the physical connectivity between two firms located in two different regions may generate exchanges of local know-how and information which increase productivity, and consequently regional performance. In other words, when the connectivity takes place at an inter-regional level, the effects it produces may also have import and export implications.

Thus, in summary, the positive effects at a firm's level are expected to reflect their beneficial influence at the level of the environment in which they perform. Thus, in economic terms, we expect a region to have the possibility to gain from network externalities, by exploiting advantages stemming from its participation in the network. At a macro-level these advantages may be indicated as the achievement of:

a) spatially dispersed information;
b) new geographic market areas;
c) complementary know-how from different specialised economic areas;
d) additional specialised input factors from other regions.

A legitimate question at this stage is under which macro-conditions the "spatial symbiosis" effect takes place. Since economic space is not homogeneous, we would expect regions each to have a different capacity to exploit network externalities. This is the subject matter of the next sub-section.

4.4.2. Macro-conditions for the exploitation of the "spatial symbiosis"

As is the case at the firm level, so also at the macro level the capability of exploiting the spatial symbiosis is not equally distributed. Essentially, a spatially varying capability of exploiting network externalities can be ascribed to two components, namely the *structural component* and the *production milieu component*.

These two elements have already been identified as the main issues for the explanation of different degrees of regional innovativeness (Davelaar, 1991). To a certain extent the same approach can be developed for the analysis of the different spatial capacity of *receptivity*, i.e. the regional capacity of "advantages appropriability". We may argue, then, that "spatial symbiosis" is highly dependent on the extent to which:

a) firms and sectors in a region are able to exploit network externalities. In other words, the regional structural component of firms and sectors with a high capacity for exploiting the "economic symbiosis" will be one of the determinants of the "spatial symbiosis". In fact, if we interpret the "spatial symbiosis" as the non-zero sum of the positive effects of network externalities on firms' performance, it is easy to understand how much the

combination of firms able to exploit network externalities in a region may generate greater "spatial symbiosis";
b) the local (spatial) capacities to exploit network externalities. In fact, the so-called "production milieu component" plays a role in the definition of the "spatial symbiosis". Identical firms, in fact, are able to exploit network externalities in different ways, according to their location in space.

The "production milieu" variables related to the development of the "spatial symbiosis" have to be looked for precisely in those spatial variables which help the achievement of a leading role in the network, and not simply the access to a network (and thus its adoption). In fact, also at a regional level the benefits a regional economy may get from joining a network depend on its role within the network. It may very well be the case that a region is in a weaker position in comparison to other regions and is exploited in its role of partner. The macro-conditions to be winner instead of loser in a network are related to the spatial elements which facilitate an innovative use of these technological potentialities.

The possibility to get more information may be analysed and exploited for the achievement of a better economic position. However, it can also be the case that the firm is unable to exploit this information for its own business. Moreover, on the contrary, it may happen that the information the firm provides to its partners is of strategic importance. In this case, that particular firm would inevitably act as a loser in the whole network.

The spatial conditions helping firms to be winners in a network are expected to be:

a) agglomeration of different innovative firms in the same region. Collective learning processes are thus generated, which enhance the capacities to exploit network externalities, through learning processes in the use and exploitation of these technologies;
b) the presence of a specialised local service sector. Instead of stressing only the role of the information infrastructure, in the case of the exploitation of the new telecommunications networks and services the existence of service firms is also crucial. Corporate organisational consultancies are essential for the integration of new telecommunications technologies with the organisational structure of firms, and - at the territorial level - with the "*vocations industrielles*" of the area;
c) an advanced labour market, in terms of technical, managerial and organisational know-how and capacity to innovate.

Seen from a static perspective, spatial symbiosis would be exploited by central areas, which are more likely to be the seedbed areas for the location of innovative firms (structural component), and also by more local economic characteristics of these firms which enhance their capacities to exploit network externalities (production milieu component). However, a dynamic analysis would point out that these conditions may, in the long run, be more in favour of peripheral regions.

4.5. Benefits and costs of network externalities

Up to now the analysis has focused on positive network externalities, by showing the net benefits firms and regions may get from joining a network, or by having a new member joining their network. A deeper interpretation of network externality effects may be interesting. The "economic and spatial symbiosis" approach rests on the assumption that the higher the degree of (physical) connectivity of firms and regions, the higher are the benefits for connected actors.

Two remarks are crucial in this respect. First of all, the marginal benefits curve is not a constantly increasing function of the number of subscribers in the network. A more realistic interpretation of this phenomenon is that the marginal utility an existing subscriber gets from a new subscriber is an increasing function for the first group of subscribers. After a certain number of subscribers it is realistic to expect that the marginal utility becomes a increasing function at decreasing marginal rates (see Figure 4.5). In fact, then the advantage a new firm may provide for a very large network is very low. Secondly, joining a network implies some costs that have to be borne by the newcomer and by the members. These are both fixed costs (in financial terms, purchase costs of the necessary equipment) and variable costs (in financial terms, price and tariff of communication). The marginal cost curve is an increasing function of the number of subscribers, and thus of the degree of connectivity (see Figure 4.5).

If we take costs of networking into account, we may also explain why, if by definition network externalities create a positive effect on the performance of firms and regions, there may still be non-connected firms.

A more precise definition of costs and benefits of network externalities is required at this stage. We may define benefits and costs as being pecuniary or non-pecuniary (see Table 4.2).

Pecuniary benefits are the ones we discussed in Section 4.3 and 4.4, i.e. network synergies among firms and the achievement of more widespread information, which respectively give rise to better economic and regional performance. Also non-pecuniary benefits arise in the presence of network externalities. At the firm level these benefits may be envisaged, for example, in the "complementary assets" a firm may get from another networked firm. It may very well be that the strategic know-how of one firm is complementary to the output of another firm and the existence of a physical connectivity may facilitate the exchange of complementary assets.

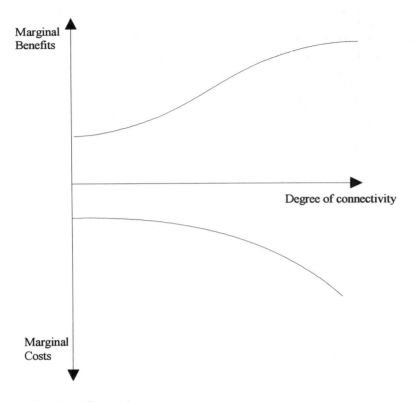

Figure 4.5. Benefits and costs curves

At a regional level, non-pecuniary benefits stemming from network externalities are the positive spillover effects that networked firms may generate at the local economy level. It may very well be the case that cumulative effects *à la* Myrdal (1959) generated by dynamic and growing firms take place which dominate regional economic growth. However, this is the only positive aspect of the network externality phenomenon at the regional level.

Costs have to be borne once a firm joins a network, which go far beyond the simple financial costs and are the negative effects of network externalities. At the firm level, pecuniary costs may be envisaged in congestion effects. For a given network, the entrance of a new subscriber in the network may generate negative network externalities, since the marginal user may create congestion on the lines. The profitability of each already existing subscriber then inevitably decreases.

Non-pecuniary costs have to be borne by a firm joining a network, since once it has become a member, some information which was previously private, has then inevitably to be shared among members. This is true in the case of the flower auction network in the Netherlands (mentioned already in Section 4.3.2) where the benefits for its members to join an international flower network, and thus an international database were diluted by the disadvantages produced by the loss of a monopolistic position with regard to strategic information.

Table 4.2. Typology of benefits and costs of network externalities

	NON-PECUNIARY	PECUNIARY
BENEFITS Firm Level Regional Level	Complementary assets Local spillover effects	Network synergies More widespread information
COSTS Firm Level Regional Level	Share of private information and know-how Share of local resources Spillover effects somewhere else	Congestion Worse regional performance via economic exploitation

Also at the regional level some negative externalities may arise while joining a network. Pecuniary costs may be incurred from the exploitation of local resources and inputs, with no direct or indirect compensation for the damage.

Moreover, some non-pecuniary costs have to be faced at the regional level. These may materialise in two kinds of costs: a) the share of local resources and b) spillover effects generated in a region which take place somewhere else. In this case, the networked region would bear the costs of the networking leaving to other local economies the possibility of exploiting the positive effects. This also holds for other kinds of infrastructure, such as the transport network (Bruinsma et al., 1989). If a region with a weak economy is provided with a network, surrounding regions might benefit relatively more. The region in question might suffer strong competition from enterprises based in more distant regions. The target region must, therefore, already have a favourable potential for economic development.

These remarks can be linked back to the traditional literature on telecommunications networks and regional development (Gillespie et al, 1989; Gillespie, 1991). In this literature the crucial question is whether backward and weak economies gain from being in a network. Our approach would formulate this question in a slightly different way, by asking whether for weak and backward economies the marginal benefits are higher or lower than the marginal costs. If marginal benefits are higher (lower) than marginal costs, a net gain (loss) is envisaged.

The economic and spatial symbiosis approach is developed according to the idea that under certain micro- and macro-conditions marginal benefits are higher than marginal costs, and thus that firms and regions may gain from entering the network. This is of course something which has to be proved empirically, and this is the subject matter of the second part of this book.

4.6. Conclusions

Our basic research questions are related to the linkage between network externalities and industrial and regional performance. In particular, our main task is the analysis of the question whether or not network externalities may be measured in terms of corporate (i.e. micro) and regional (i.e. macro) performance. The conceptual framework presented above is the first attempt towards the conceptualisation of the role network externalities play in the performance of firms and regions.

As we have underlined, our framework suggests that a set of polarisation effects arises via the physical linkage of firms. This is not a new concept, since the analyses developed by Perroux and Boudeville on the concepts of "growth pole" and "growth centres", respectively, in the middle of the century were rather similar in terms of their respective explanation of the effects that synergic and cumulative attraction factors could cause at a firm and regional level. What is rather new in our approach is the *nature* of these effects, which is profoundly different from that postulated by the traditional theories of Perroux and Boudeville. The causes explaining the cumulative effects on the corporate and regional performance are no longer attributed to key production units able to generate attraction factors. Instead, our framework explains the causes of the cumulative effects through the existence of a physical connection with other firms, which are therefore able to generate strong synergies reflected in their productivity function, and thus on their corporate and aggregate performance.

The "economic and spatial symbiosis" framework represents a coherent and logical model of how network externalities impact on the performance of firms and regions. However, at a purely conceptual level, some considerations have been made with regard to the fact that the simple adoption of information and communication technologies does not necessarily mean economic advantages for firms and regions. As we have underlined in this chapter, some benefits but also costs may arise by joining the network and a typology of these benefits and costs has been provided. The existence of costs in joining a network may explain why firms do not enter a network, although, by definition in our conceptual framework, network externalities have a positive impact on the performance of firms. The micro- and macro-conditions to be winners or losers in a network are mentioned, and their validity will be demonstrated empirically for the Italian case.

In order to show the relevance of the "economic and spatial symbiosis" model, it is now necessary to develop a framework of analysis that is able to provide a testable set of hypotheses regarding network externality effects on both the performance of firms and regions. This is the subject matter of the next chapter.

Notes

1. See Karshenas and Stoneman, 1990.
2. See Section 4.5 for a precise definition of benefits and costs of network externalities.

5 A theoretical model for network externality analysis

5.1. Introduction

In the previous chapter a conceptual framework was presented, which is built on some crucial assumptions. The general idea underlying this framework is that network externalities are one of the most important reasons for entering the network. The second idea is that there is a strong linkage between network behaviour and industrial and regional performance. In other words, the fundamental assumption is that there is a linkage between the number of existing members in the network and the industrial performance of the firm. The same linkage is expected at a regional level, where the aggregate production function is influenced by the number of firms connected to the network. The advantages obtained by these firms exert a positive cumulative effect on regional development. At a conceptual level this framework represents a coherent and consistent model of how network externalities work on the performance of firms and regions, but, nevertheless, its ultimate validity depends on the possibility of testing it in an empirical context.

In this chapter we present a theoretical model for network externality analysis. In order to prove the validity of the "economic and spatial symbiosis" framework, it is important to find a methodology able to analyse and measure the effect of network externalities on industrial and regional performance. Our idea is to develop a theoretical model to act as a *bridge* between our conceptual assumptions and our empirical analysis. The aim of the model presented in this chapter is not to perform a statistical analysis; its relevance is rather focused on the identification of appropriate *testable propositions*, which can be validated later by means of a suitable data set. In fact, from this theoretical mathematical model, some crucial propositions are highlighted, which have to be proved empirically. In other words, its nature is essentially qualitative.

The model deals with a situation of cost minimisation in the presence of network externalities affecting the production function by means of typical technical externalities. The chapter will first describe the model in general terms, underlining our mathematical modelling of the externality phenomenon (Section 5.2). We then move on to a more focused analysis developed in both the case of symmetric and asymmetric firms acting in the market. The model is developed first under the assumption of two firms (Section 5.3), and will then be generalised for n firms (Section 5.4). The results of the two-agents case will be presented in a summary section following a game-theoretic approach. Section 5.5 relates the results to the case of telecommunications networks, while Section 5.6 contains the limits of the model. Finally, Section 5.7 presents the qualitative propositions resulting from our theoretical approach. These propositions need to be tested empirically in order to prove the validity of our "economic and spatial symbiosis" framework. Thus, this chapter is an introduction to the empirical part, which will be presented in the second part of this book.

5.2. Network externalities and micro-economic production theory

5.2.1. The "technology effect" and the "network effect"

Network externalities have generally been analysed as consumption externalities, acting on the utility function of an individual and generating interdependent decision-making processes. As we anticipated in Chapter 2 and explained in Chapter 4, our approach in this study is rather a new one, in which we underline that network externalities may also influence the production function, acting as *technical production externalities*. From this perspective, network externalities may be defined as follows. Let:

$$Y_i = Y_i(K_i, L_i, N_{i1})$$

be the production function of firm i, characterised by a certain amount of capital (K), of labour (L), and of a certain volume of information (N). The volume of information (N) is dependent on two variables: on the one hand, it depends on the kind of technology (T_{i1}) which characterises a specific network (network 1) to which the firm i is linked, and on the other on the number of subscribers (S_1) that are linked to network 1. Thus the volume of information for firm i reads as:

$$N_{i1} = N_{i1}(T_{i1}, S_1)$$

where T_{i1} is the technology of network 1 to which firm i is linked and S_1 is the number of subscribers using network 1. The technology (T) is influenced by any (endogenous or exogenous) technological change. There are positive network externalities if:

$$Y_i(K_i, L_i, N_{i1}) > Y_i(K_i, L_i, N'_{i1})$$

where $N'_{i1}(T_{i1}; S_1 + 1) > N_{i1}(T_{i1}; S_1)$

This means, in fact, that the advantages of a greater volume of inputs obtained via the network (dependent on the number of subscribers (or firms) connected to the network), has, *ceteris paribus,* positive effects on the production function of a firm. It is evident that the higher the number of firms using the network, the greater the advantages for firm i.

This definition is useful for another clarification of the concept of production network externalities, as it explains the distinction between the general effects of a technological innovation on the production function and the specific effects of network externalities. At a first glance one could in fact argue that the increase in the productivity of firms can be the result of the adoption of advanced technologies, which allow the achievement of greater efficiency, as would happen in the case of any innovation introduced in a firm. Network externalities generate the same effects as technological innovation, but the *nature* of these effects is entirely different.

Let T_{i1} and T_{i2} be two different technologies, which characterise respectively an old (1) and a new network (2) (e.g. the telex and the teletex networks). Let us assume that every firm is either linked to the old or to the new network, where the possibility of a transition to a third network is excluded. The volume of information included in the production function may come from the link either with the network characterised by the old technology (N_{i1}), or by the new technology (N_{i2}). In this case network externalities occur if:

$$Y_i(K_i, L_i, N_{i1}) < Y_i(K_i, L_i, N'_{i1})$$

where $N'_{i1}(T_{i1}; S_1 + 1) > N_{i1}(T_{i1}; S_1)$

and if

$$Y_i(K_i, L_i, N_{i2}) < Y_i(K_i, L_i, N'_{i2})$$

where $N'_{i2}(T_{i2}; S_1 + 1) > N_{i2}(T_{i2}; S_1)$

A firm chooses to switch from the old network characterised by the technology T_1 to the new network, with the technology T_2, if:

$$Y_i(K_i, L_i, N_{i2}) - Y_i(K_i, L_i, N_{i1}) > 0$$

where $N_{i2}(T_{i2}; S_2 + 1) > N_{i1}(T_{i1}; S_1)$

This inequality shows that firm i enjoys two effects, the *technology effect*, which represents the increase of productivity generated by the new technology (T_{i2}), and the *network effect*, which determines the difference in the volume of information obtained by a greater number of subscribers linked to the new network ($S_2 + 1$) compared with the lower number of subscribers linked to the old network (S_1)[1].

The "economic and spatial symbiosis" framework presented in Chapter 4 is based on the *network effect*. Our aim in the next chapters is to find a way of separating the network effects from the technology effects, so as to be able to measure the network effect, despite the technology effect.

5.2.2. A network externality model

Traditional micro-economics has developed a solid model to explain the firm's behaviour as far as resource allocation problems are concerned. Under the assumption of rational behaviour influencing decision-making processes, a firm will be inclined to allocate its resources in such a way so that its marginal costs are equal to its marginal benefits, thus maximising its profits. This also means that the firm's equilibrium solution is represented by the intersection point between the production function (isoquant) and the cost function (isocost). This solution, as is well known, guarantees the best resource allocation for a given output.

Our main question at this point is to see what happens to the equilibrium solution in the presence of network externalities, i.e. in the presence of *advantages in the production function stemming from the connection with other firms*.

Let us assume the existence of a general production function of firm i:

$$Y_i = Y_i(K_i, L_i, N_i)$$

where K_i and L_i are the traditional production factors capital and labour, and N_i represents *the volume of information* obtained via the network, thus becoming a third production factor for firm i. The volume of information obtained by a firm depends on the *number of contacts* firm i has with firm j (obtained as the sum of contacts useful for firm i, generated either by firm i or by firm j). The contribution to the aggregate volume of information of being connected with firm i is described by the following *network connectivity function*:

$$N_i = N_i(I_i, \Sigma_j I_{ji})$$

where I_i denotes the number of contacts firm i generates with other firms and $\Sigma_j I_{ji}$ represents the number of contacts that other firms generate towards firm i. This function contains network externality effects, since the higher the number of contact generated by other firms, the lower the number of contacts firm i has to generate towards other firms in order to achieve the same volume of information.

The costs of contacts involved for firm i consist of two parts, i.e. the costs related to the purchase of the necessary equipment (*equipment costs*) and the costs related to the use of the network (*use costs*).

The first part, *equipment costs*, can be represented by subtracting financial resources used to buy this equipment ($c_{ni} N_i$) from the total amount of financial resources of the firm. c_{ni} represents the unit cost of information in terms of telecommunications equipment purchase and N_i the volume of information obtained by the network. The equipment costs depend on the volume of information obtained by the firm, since one may reasonably expect that after a certain volume of information run on the network, a firm is obliged to invest in new equipment (new networks) because of the technical constraints of the old network.

The second category of costs, i.e. *use costs*, can be added as a separate component in the cost equation and be thought of as a function of the number of contacts:

$$C_{u_i} = c_{I_i} I_i$$

where C_{ui} is the total use cost of telecommunications services for firm i, c_{Ii} is the unit cost of use for firm i and I_i is the number of contacts generated by firm i.

In particular, let us now assume the case of fixed output for firm i. Minimisation of costs, given the output level Y^*, is now supposed to be the objective of firm i; this leads to the following first order conditions:

$$\delta Y_i / \delta L_i = c_{L_i} / \lambda$$

$$\delta Y_i / \delta K_i = c_{K_i} / \lambda$$

$$(\delta Y_i / \delta N_i)(\delta N_i / \delta I_i) = (c_{I_i} / \lambda) + c_{n_i}(\delta Y_i / \delta K_i)(\delta N_i / \delta I_i)$$

where c_{Li} and c_{Ki} represent the unit costs, respectively, of capital and labour. In the case of labour and capital these conditions indicate the usual equality of marginal productivity and marginal costs (price of production factors). For the "information" production factor a similar result is obtained, but in this case the marginal costs include the *opportunity costs* of capital, i.e. the costs of using financial resources for the purchase of telecommunications equipment, in terms of other fixed capital which could otherwise be purchased[2].

In comparison with the case where network externalities are not present, we would expect:

a) firm i to achieve the same production level with lower quantities of capital and labour involved in the production process. This expectation is derived from two considerations. First of all, information obtained via the network can substitute traditional production factors. Secondly, capital and labour can become more productive in the presence of more information and more know-how (productivity effect). The effects can also be shown with the help of a chart (Figure 5.1), where the production function Y representing a fixed output is shifted towards the origin. In fact, the same production level represented on a new isoquant Y' can be achieved with the use of lower capital and labour. This chart is exactly the same as the "technological change" chart of the neoclassical Hicks model. However, one important difference can be underlined; the shift of the isoquant towards the origin and the subsequent reduced levels of capital and labour does not only depend on the behaviour of that particular firm, but also on what the other firms are doing. Paradoxically, in the presence of network externalities, both the investment behaviour of firm i and the investment behaviour of other firms are strategic to the shift of the isoquant curve of firm i;

b) firm i to incur lower production costs, the higher the number of firms linked to the network. In the presence of a high number of firms connected, the use costs would be divided among a greater number of people, with a subsequent decrease of production costs for the connected firms.

In the rest of the chapter we will present a model in order to arrive at some mathematical results and see whether our expectations concerning the behaviour of firms in the presence of network externalities are plausible. If they are, we are in a position to derive some crucial propositions to be investigated in the empirical analysis.

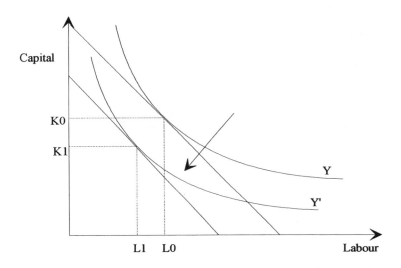

Figure 5.1. Effects of network externalities on the production function

The production function is expressed in the form of a Cobb-Douglas function. The choice of the Cobb-Douglas production function is plausible as a general approach, since it is based on the principle of substitutability among production factors, with an elasticity of substitution equal to 1. These features inevitably influence the results. In fact, with a Leontief type of production function, one might expect different results, since in this case production factors are non-substitutable. On this issue one can easily argue that in the real world production is based on the choice of scarce resources (i.e. substitutional) and for this reason the Leontief case represents the exception rather than the rule.

While it is easy to justify the choice between Leontief and Cobb-Douglas production functions, one might easily argue that there are other kinds of production functions that show a certain degree of substitution among production factors. Especially CES (constant elasticity of substitution) production functions, which are generalised functions containing Cobb-Douglas and linear functions according to the values assumed by the parameters[3], are more comprehensive and thus realistic, although more difficult to handle. With these kinds of production functions we could have had different degrees of substitution than in the case of the Cobb-Douglas; with Cobb-Douglas production functions the elasticity of substitution assumes a value 1. The easy mathematical characteristics of Cobb-Douglas, leading to analytical solutions, have rather influenced the choice, but the previous considerations should be borne in mind when the results are analysed.

The model has been developed under the assumption of a given output. This choice has been made because the database we use to develop the empirical analysis does not allow dynamic studies.

The analysis is developed first in the case of two firms, under the hypotheses of a competitive situation (Section 5.3.1) and of a leader/follower situation (Section 5.3.2.). The analysis is then applied to the case of n firms (Section 5.4.). Although the mathematical model has some intrinsic limits, it is nevertheless very useful to arrive at some propositions (Section 5.5.) which will be tested empirically in the following chapters.

5.3. A specific network externality model

5.3.1. A simultaneous game with two symmetric agents

In this section we consider the existence of two symmetric firms facing the choice of an optimal resource allocation among the two traditional production factors, capital and labour, and new inputs such as information and know-how obtained by being linked to a telecommunications network. The volume of these inputs depends on the number of firms connected to that particular network. The network externality concept is present in our model by assuming that the volume of information firm 1 receives is dependent on the degree of connectivity with other firms. The volume of inputs via the network does not only depend on the contacts generated by firm 1, but also on the contacts made by firm 2 with firm 1.

The *production function* of firm 1 reads as:

$$q_1 = \eta K_1^{\alpha_1} L_1^{\beta_1} N_1^{\delta_1} \qquad \text{with } \alpha_1 + \beta_1 + \delta_1 = 1$$
$$\text{and } \alpha_1, \beta_1, \delta_1 > 0$$

where q represents the output of firm i[4], K and L represent the traditional production factors, capital and labour, while N represents the volume of information and know-how obtained via the network, dependent on the number of contacts undertaken by firm 1 and by firm 2.

We assume a linear relation between the number of contacts on a network and the volume of inputs obtained via the network. Thus, the following linear equation in N, represents the *network connectivity function*:

$$N_1 = \varepsilon_{11} I_1 + \varepsilon_{21} I_2$$

where I_1 represents the number of contacts generated by firm 1 towards firm 2, and I_2 represents the number of contacts generated by firm 2 towards firm 1. ε represents the *efficiency of the contacts* in terms of the volume of information. In particular, ε_{11} represents the efficiency of contacts for firm 1, generated by firm 1, while ε_{21} represents the efficiency of contacts for firm 1, generated by firm 2. It is in fact plausible to think that not all contacts are strategic for the firm and enter the production functions. Some contacts may contain some relevant information as well as already known information. Only a part of the contact will be of interest to the firm and will be added to the volume of information strategic to the firm. For the same reason, we expect ε_{11} to be greater than ε_{21}, since the contacts generated by the interested firm have a higher probability of being more relevant for firm 1.

The *networking cost function* in our model is dependent on two elements: the equipment costs and the use costs. The first category is a function of the volume of information and is obtained by subtracting from the total amount of financial resources of firm 1 that part of financial resources devoted to telecommunications equipment purchase ($c_{n_1}N_1$), where c_{n_1} is the unit cost per information of telecommunications equipment and N_1 the volume of information required by the firm. The second category of costs is related to the use of the network and is of course dependent on the number of contacts generated by firm 1:

$$C_{u_1} = c_{I_1} I_1$$

The interesting question at this point is to find the optimal levels of inputs (K, L, and I), guaranteeing the minimisation of costs given the output level q*. The problem for firm 1 is thus:

$$\min C_{tot_1} = c_{K_1} K_1 + c_{L_1} L_1 + c_{n_1} N_1 + c_{I_1} I_1$$

$$\text{s.t.} \quad q_1^* - \eta K_1^{\alpha_1} L_1^{\beta_1} N_1^{\delta_1} = 0$$

$$N_1 - \varepsilon_{11} I_1 - \varepsilon_{21} I_2 = 0$$

The solution is achieved by solving the following Lagrangian function:

$$\mathcal{L}_1 = c_{K_1} K_1 + c_{L_1} L_1 + c_{n_1} N_1 + c_{I_1} I_i + \lambda_1 \left(q_1^* - \eta K_1^{\alpha_1} L_1^{\beta_1} N_1^{\delta_1} \right) + \lambda_2 \left(N_1 - \varepsilon_{11} I_1 - \varepsilon_{21} I_2 \right)$$

First order conditions of cost minimisation imply:

$$K_1 = \frac{c_{L_1} \beta_1 L_1}{c_{K_1} \alpha_1}$$

$$N_1 = \frac{\varepsilon_{11} \delta_1 K_1}{\alpha_1} \left(\frac{1}{\frac{c_{I_1}}{c_{K_1}} + c_{m_1} \varepsilon_{11}} \right)$$

Solving the system above, we obtain the demand functions for the production factors that guarantee the minimisation of costs, given a fixed output[5]. The demand function for the number of contacts of firm 1 is the following[6]:

$$I_1 = \frac{q_1^*}{\varepsilon_{11} \eta_1} \left(\frac{c_{L_1} \beta_1}{c_{K_1} \alpha_1} \right)^{\beta_1} \left(\frac{\delta_1}{\frac{\alpha_1 c_{I_1}}{\varepsilon_{11} c_{I_1}} + \alpha_1 c_{m_1}} \right)^{\alpha_1 + \beta_1} - \frac{\varepsilon_{21}}{\varepsilon_{11}} I_2$$

The difference from the traditional model of production factors allocation, given a certain level of output, is that the volume of information does not only depend on the choices of firm 1, but also on what firm 2 decides. In fact, the volume of information for firm 1 also depends on the number of contacts of firm 2. The present I_1 is found independently of the choice of firm 2. This means that firm 1 minimises its production costs without taking into consideration the strategic choices of firm 2. Since firm 2 behaves rationally, its aim will also be the minimisation of costs and, thus, it will choose the number of contacts in the network (I_2) until the level guarantees a profit maximisation. If firm 2 changes the level of I, this decision influences the volume of information. In this case, firm 1 will be obliged to reallocate its production factors until the level of I_1 guarantees a simultaneous cost minimisation for both firm 1 and 2.

This is a typical *Cournot equilibrium model*, where the equilibrium solution is found at the intersection point between the two reaction functions, i.e. at the point where the two firms simultaneously achieve a cost minimisation. The distance to the minimum level of I involved for one of the two firms could stimulate the firm to reallocate its production factors when seeking a cost minimisation, and this would oblige the other firm to reallocate its production factors. The game finds a stable solution only if both firms simultaneously achieve a cost minimisation.

If the production function, the network connectivity function and the cost function have the structure designed above, the above model has the following properties:

1. cross-price elasticities of labour and volume of information are positive: the volume of information increases when the cost of labour increases and vice versa. As expected, there is a non-zero *substitution* between labour and volume of information. The same cannot be said in the case of capital and volume of information. These two variables are inter-related via the "opportunity cost", thus when the cost of fixed capital increases, it is not immediately the case that the volume of information increases;
2. the increase in the unit cost of the number of contacts of firm 1 leads to a decrease in the number of contacts generated by firm 1;
3. the increase in the number of contacts generated by firm 1 leads to a decrease in the number of contacts generated by firm 2;
4. the increase in the efficiency of the contacts generated by firm 2 for firm 1 (ε_{21}) leads to a decrease in the number of contacts generated by firm 1;
5. the effect of an increase in the efficiency of the contacts generated by firm 1 (ε_{11}) on the number of contacts leads to a decrease in the number of contacts;
6. the increase in the unit cost of equipment (c_{ni}) of firm 1 leads to a decrease in the number of contacts of firm 1.

All these results are plausible and realistic. One of the main conclusions concerning the above model is that there is a degree of substitution between labour and volume of information. This is of course related to the specific form of the production function assumed: Cobb-Douglas production functions imply a high degree of substitutability. The possibility of substitution among production factors in the presence of network externalities means that the higher the effort made by firm 2 to be connected, the higher the advantages firm 1 achieves on its production function in terms of decrease in production costs. For this reason it is interesting to see what the results are in the presence of asymmetric agents on the market. This is the subject matter of the next section.

5.3.2. *A sequential game with two asymmetric agents*

In this section we remove the assumption of two symmetric agents in the market. The problem is to see whether the equilibrium point changes under the assumption that a firm is in a leading position in the market. In particular, we recall the *Von Stackelberg model*, assuming the existence of a leading firm in the market, and of a "follower". An example of this is the case of IBM in the software market, leading the market with a turnover of approximately 11,000 million US $ in 1991, and followed by a high number of smaller firms[7]. It is a two stage procedure, where the decisions are made in the first stage by the leader, who aims at profit maximisation. This solution is imposed on the follower, who will take it as given and will maximise its profits under the conditions decided by the leader.

In our model let us suppose that firm 1 is the leader and firm 2 the follower. The problem of the follower is to minimise its costs for a given output:

$$\min C_{tot_2} = c_{K_2} K_2 + c_{L_2} L_2 + c_{n_2} N_2 + c_{I_2} I_2$$

$$\text{s.t.} \quad q_2^* - \eta K_2^{\alpha_2} L_2^{\beta_2} N_2^{\delta_2} = 0$$

$$N_2 - \varepsilon_{12} I_1 - \varepsilon_{22} I_2 = 0$$

where the costs of the follower depend on the level of I_1 decided by the leader. This value is given for the follower, since it is decided by the leader in a previous stage. Firm 2 will then determine the level of I_2 for a given I_1.

On the contrary, the problem of the leader is to achieve a cost minimisation taking into account what the reaction of the follower will be. The leader is conscious that its decisions will influence the action of the follower. For this reason the leader will take into account the reaction of the follower. The follower has to minimise its costs, taking into account the (minimum) level of investments for firm 2. The leader problem will thus be:

$$\min C_{tot_1} = c_{K_1} K_1 + c_{L_1} L_1 + c_{n_1} N_1 + c_{I_1} I_1$$

$$\text{s.t.} \quad q_1^* - \eta K_1^{\alpha_1} L_1^{\beta_1} N_1^{\delta_1} = 0$$

$$N_1 - \varepsilon_{11} I_1 - \varepsilon_{21} I_2 = 0$$

where I_2:

$$I_2 = \frac{q_2^*}{\varepsilon_{22}\eta_2}\left(\frac{c_{L_2}\beta_2}{c_{K_2}\alpha_2}\right)^{\beta_2}\left(\frac{\delta_2}{\dfrac{\alpha_2 c_{I_2}}{\varepsilon_{22}c_{I_2}}+\alpha_2 c_{n_2}}\right)^{\alpha_2+\beta_2}\frac{\varepsilon_{12}}{\varepsilon_{22}}I_1$$

The problem is solved with the following Lagrangian function:

$$\mathcal{L}_1 = c_{K_1}K_1 + c_{L_1}L_1 + c_{n_1}N_1 + c_{I_1}I_1 + \lambda_1\left(q_1^* - \eta K_1^{\alpha_1} L_1^{\beta_1} N_1^{\delta_1}\right) + \lambda_2\left(N_1 - \varepsilon_{11}I_1 - \varepsilon_{21}I_2\right)$$

The first order conditions for cost minimisation imply:

$$K_1 = \frac{c_{L_1}\beta_1 L_1}{c_{K_1}\alpha_1}$$

$$N_1 = \left(\varepsilon_{11} - \frac{\varepsilon_{21}\varepsilon_{12}}{\varepsilon_{22}}\right)\frac{\delta_1 K_1}{\alpha_1}\frac{1}{\dfrac{c_{I_1}}{c_{K_1}} + \left(\varepsilon_{11} - \dfrac{\varepsilon_{12}\varepsilon_{21}}{\varepsilon_{22}}\right)c_{n_1}}$$

The demand equation for the number of contacts is obtained by solving the above system (see annex 1 at the end of the book), and the result is represented by the following equation:

$$I_1 = \frac{1}{\varepsilon_{11} - \dfrac{\varepsilon_{21}\varepsilon_{12}}{\varepsilon_{22}}} \frac{q_1^*}{\eta_1}\left(\frac{c_{L_1}\beta_1}{c_{K_1}\alpha_1}\right)^{\beta_1}\left(\frac{\delta_1}{\dfrac{\alpha_1 c_{I_1}}{\left(\varepsilon_{11}-\dfrac{\varepsilon_{21}\varepsilon_{12}}{\varepsilon_{22}}\right)c_{K_1}}+\alpha_1 c_{n_1}}\right)^{\alpha_1+\beta_1} - \frac{\varepsilon_{21}}{\left(\varepsilon_{11}-\dfrac{\varepsilon_{21}\varepsilon_{12}}{\varepsilon_{22}}\right)}\cdot A$$

where $A = \dfrac{q_2^*}{\varepsilon_{22}\eta_2} \left(\dfrac{c_{L_2}\beta_2}{c_{K_2}\alpha_2}\right)^{\beta_2} \left(\dfrac{\delta_2}{\dfrac{\alpha_2 c_{I_2}}{\varepsilon_{22} c_{I_2}} + \alpha_2 c_{n_2}}\right)^{\alpha_2+\beta_2}$

In the case of two asymmetric agents in the market, the following considerations can be drawn from our analysis:

1. the increase in the labour cost of the follower creates an increase in the number of contacts for the leader, since the number of contacts generated by the follower inevitably decreases;
2. the increase in the efficiency of contacts generated by the follower for the leader (ε_{21}) generates a decrease in the number of contacts generated by the follower, and thus the number of contacts generated by the leader increases, although it increases less than in the case of two symmetric agents;
3. the volume of information used in the production function of the leader is lower than the volume of information used by the firm in a symmetric market situation.

We will now attempt to generalise our model for the case of n firms. Before doing this, in the next section we provide a summary of the present results in the form of a game theory approach.

5.3.3. Synthesis of results: a game-theoretic approach

In the previous section we investigated the possibility of finding equilibrium solutions in the presence of two firms, developing the model by analysing two different market situations. In this summary of our results, we would now like to take into account the different decisions that can be made by the two firms, simultaneously or sequentially, through a game-theoretic application. The result we will achieve is that the equilibrium found in the previous section is not only a Nash equilibrium, i.e. two situations where the choice of one firm (firm 1) is optimal, given the choice of the other firm (firm 2) in the market, and the choice of firm 2 is optimal, given the choice of firm 1 (Nash, 1950a, 1950b and 1952), but is also a Pareto-optimal solution for the two firms, in the cases of both Cournot and Von Stackelberg oligopolistic solutions.

To achieve this result, we will present two games; in the first game, two firms exist in an oligopolistic market which are not connected but have the possibility to connect via a network. In the second game, n firms exist in an oligopolistic market, where n-1 are connected to the network and only 1 is outside the network. In this case, as we will see below, for the n-1 firms the choice is either to accept or to put obstacles in the way of the potential entrant who wishes to join the network, while for the potential entrant its choices are to enter or not to enter.

Let us assume the existence of two firms, acting in an oligopolistic market with a fixed level of capital and labour. Both firms have two possibilities in relation to the network. They can both decide either to "enter the network" or "not to connect". In the first case, they will obtain network externality advantages, in the second case they lose these advantages.

Table 5.1.
Pay-off matrix in terms of number of contacts in a two firms game on the decision to connect or not

FIRM 1

		Connect	Not Connect
FIRM 2	Connect	I; I	-I; zero
	Not Connect	zero; -I	zero; zero

I = Number of contacts undertaken by firm x

The pay-off matrix of this game is presented in Table 5.1, where the pay-off values are expressed in terms of contacts of information that two firms achieve if they are networked. In our "economic symbiosis" framework, firms gain from the connectivity because they obtain from the network more information, which acts as an additional production factor in the production function. If we assume the hypothesis of fixed capital and labour, the maximisation of profits depends on the amount of information of these firms. If firm 1 makes the decision to enter and firm 2 behaves in the same way, the advantages will be to obtain a profit maximisation under the Cournot equilibrium, and the number of contacts pursued by the two firms are the Cournot levels obtained in Section 5.3.1. In the case of sequential decisions, the number of contacts in this case will be reflected by the Von Stackelberg solution of Section 5.3.2.

If one firm does not enter, and the other behaves the same way, the advantages from being networked are equal to zero for both firms. Finally, if one firm decides to enter, and the other decides not to connect, the pay-off of the one which enters will be a loss, since the entrant pays the cost of networking, without having the benefits of being networked.

Two Nash-equilibrium game solutions exist. In fact, if firm 1 decides to enter the network, firm 2 gains more if it makes the decision to enter. This is also valid if firm 2 chooses to enter, as the greater advantages firm 1 obtains are represented by the decision to enter. However, if firm 1 decides not to enter, firm 2 has advantages by choosing the non-enter solution. This holds also for the choice of firm 2 not to enter, because firm 1 will have the greater advantages from the choice of not to enter. Both strategies ("enter the network / enter the network") and ("do not enter the network / do not enter the network") are Nash equilibrium solutions.

This is a "prisoner's dilemma" game. In fact, two Nash equilibrium points exist, but only one is a Pareto-optimality equilibrium. If one firm knew in advance the choice of the other, and under the assumption that for a firm it would be better to enter than not enter a network, the choice would be the Nash-Cournot solution which is also a Pareto-optimality point, i.e. the strategy "enter the network" / "enter the network".

The same situation is achieved in the case that there are n firms in an oligopolistic market, and that only n-1 firms are connected to a network. These firms, behaving as a single decision-maker, have two opportunities; they can either decide "to accept the entrance" in the network of the non-connected firm (potential entrant), or to "obstruct its entrance". In the first case, firms will gain also from being linked to an additional firm, in the second case this advantage will not exist. From its side, the potential entrant can choose between two possibilities, either "to enter the network" or "not to enter the network"; in the first case, it will obtain network externalities, in the second case the advantages of being networked will be equal to zero.

The pay-off matrix of this game is similar to the previous one and is presented in Table 5.2. The values represent the amount of information the potential entrants and the existing firms achieve if they are networked. These values are a measure of the advantages firms receive on the production function. If we assume the hypothesis of fixed capital and labour, the minimisation of costs (and thus the maximisation of profits) depends on the level of network connectivity of these firms. If the potential entrant enters and the existing firms accept, the advantages will be to obtain a profit maximisation under the Cournot equilibrium, and the number of contacts pursued by the entrant and by the existing firms are the Cournot levels obtained in Section 5.3.1. If we assume a sequential decision on the level of network connectivity to be achieved, the Von Stackelberg solution of Section 5.3.2. will hold.

If the potential entrant decides to enter, but the existing firms in the network obstruct this decision, the pay-off for the existing firms will be equal to zero, because they do not obtain the advantages of an additional user in the network. The new entrant will face only disadvantages, since the entrant pays the costs of networking, without having the benefits of being networked. If the potential entrant decides not to enter the network, the level of network connectivity will be zero for both the entrant and the existing firms, regardless of the decision made by the already networked firms.

Table 5.2.
Pay-off matrix in terms of number of contacts in a firms game on the decision to let one firm enter the network

N-1 FIRMS

		Connect	Obstruct
FIRM 2	Connect	I; I	-I; zero
	Obstruct	zero; -I	zero; zero

I = Number of contacts undertaken by firm x

Again two Nash equilibrium solutions exist for this pay-off matrix. In fact, if firm 1 makes the decision to enter, for the existing firms the best decision to make is to achieve a Cournot solution and thus accept the entrance. If the potential entrant decides not to enter, the existing firms have no preference for accepting or creating obstacles to the potential entrant, since their pay-off is in both cases null. If the existing firms decide to accept the entrance, for the potential firm the pay-off is higher if it makes the decision to enter, and thus accepts a Cournot solution. If the already existing firms decide to obstruct the potential firm, its the best choice is not to enter. Also in this game, the Cournot solution found in the previous sub-section is both a Nash equilibrium point and a Pareto-optimality solution, under the assumption that for the new firm it is better to enter than not to enter the network.

5.4. A generalisation of the model

5.4.1. A simultaneous game with n firms

In this part of the analysis the aim of the work is to extend our model to the case of n firms. In particular, we deal with the case of 3 firms. What holds for

3 firms can in fact be generalised for n firms, as indirect links can also be taken into account.

These firms can be linked together in different ways, i.e. a direct linkage among the three can exist, and also an indirect link can characterise their relationship. We develop the model in the case of three different kinds of linkages existing among our three firms, namely (Figure 5.2):

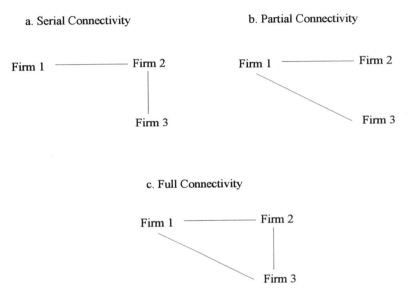

Figure 5.2. Possible connectivity among firms

- the case of *serial connectivity*, i.e. the case of firm 1 being connected with firm 2 directly, and firm 2 with firm 3, but where a direct linkage between firms 1 and 3 does not exist. In this case we expect firm 1 to gain from network externalities generated by the link between firm 1 and 2, and to gain in *a very limited way* by the link between firms 2 and 3, in comparison to the following case;
- the case of *partial connectivity*, i.e. the case of firm 1 directly linked to both firms 2 and 3 but where there is no direct link between firms 2 and 3. The advantages firm 1 gets from being networked are greater than in the case of serial connectivity, since the contacts are all directly linked to firm 1;

- the case of *full connectivity*, i.e. the case of n firms being fully connected. This is the case where network externality advantages are greater, since the number of connected people is higher.

Some considerations will be presented for the case of an asymmetric game, although the analysis in the case of n firms is again restricted to the case of a leader in the market, and n-1 Cournot competitors, as we will see in Section 5.4.2.

Serial connectivity. Let us start with the case of three firms, where firm 1 is directly connected with firm 2 and firm 2 with firm 3, but where the link between firms 2 and 3 is missing. Firms have the same role in the market, i.e. none of them is in a leading position. Firms have also similar costs and production functions.

Information firms get via the network is a strategic production factor. The volume of information firm 1 gets is dependent not only on the number of contacts between firm 1 and firm 2, but, because firm 2 is linked to firm 3, this link will influence the volume of information firm 1 gets, although in a limited way with respect to the case where the link between firm 1 and firm 3 is direct. The network externality concept is present by assuming that the volume of information firm 1 receives is dependent on the number of contacts with other firms. In our specific case, network connectivity depends not only on the number of contacts that firm 1 generates to be linked with firm 2 (and vice versa), but also on the number of contacts firms 2 and 3 have between them.

The *production function* has again a Cobb-Douglas structure, namely for firm 1:

$$q_1 = \eta K_1^{\alpha_1} L_1^{\beta_1} N_1^{\delta_1} \quad \text{with } \alpha_1 + \beta_1 + \delta_1 = 1$$
$$\text{and } \alpha_1, \beta_1, \delta_1 > 0$$

where K represents the capital stock of firm 1
L represents the labour of firm 1
N represents the network connectivity (or the volume of information) of firm 1

In general, N is a function of how many contacts firms 1 and 2 have, which depend on the number of calls firm 1 makes to firms 2 ($I_{1,2}$), and on the number of calls which it receives from firms 2 ($I_{2,1}$); and how many contacts firms 2 and 3 entertain among themselves ($I_{2,3}$). The distinction between who calls and who receives the call is not interesting in this second case, since these costs are in any case not related to firm 1. The assumption of a linear relation between the number of contacts in the network and the volume of information still holds, and the *network connectivity function* is thus:

$$N_1 = \varepsilon_{1,2} I_{1,2} + \varepsilon_{2,1} I_{2,1} + \tau_{2,3}$$

For the above relation the assumption that the efficiency of the contacts firms 1 and 2 entertain between themselves ($\varepsilon_{1,2}$ and $\varepsilon_{2,1}$) is lower than the efficiency of the contacts between firm 2 and 3 (τ) because what firm 1 gets from the linkage between firms 2 and 3 is a very limited amount in respect to the immediate advantages of being linked directly with firm 2.

The *cost function of networking* for firm 1 can be separated into two categories: the cost of use ($c_{I_{1,2}} I_{1,2}$) and the costs of purchase of the equipment ($c_{n_1} N_1$).

The problem is again a minimisation of costs for firm 1 given a certain level of fixed volume of output, solved with the use of a Lagrangian function subject to a certain level of fixed output:

$$\mathcal{L}_1 = c_{K_1} K_1 + c_{L_1} L_1 + c_{n_1} N_1 + c_{I_{1,2}} I_{1,2} + \lambda_1 \left(q_1^* - \eta K_1^{\alpha_1} L_1^{\beta_1} N_1^{\delta_1} \right) + \lambda_2 \left(N_1 - \varepsilon_{1,2} I_{1,2} - \varepsilon_{2,1} I_{2,1} - \tau_{2,3} \right)$$

The solution is given by finding the first order conditions for K, L and I. In order to minimise costs, the demand function of the number of contacts for firm 1 is the following[8]:

$$I_{1,2} = \frac{q_1^*}{\varepsilon_{1,2} \eta} \left(\frac{c_{L_1} \beta_1}{c_{K_1} \alpha_1} \right)^{\beta_1} \left(\frac{\delta_1}{\frac{\alpha_1 c_{I_1}}{\varepsilon_{1,2} c_{I_1}} + \alpha_1 c_{n_1}} \right)^{\alpha_1 + \beta_1} - \frac{\tau}{\varepsilon_{1,2}} I_{2,3} - \frac{\varepsilon_{2,1}}{\varepsilon_{1,2}} I_{2,1}$$

This leads to the following effects:

1. the increase in the costs of labour generates an increase in the number of contacts generated by the firms in order to increase the third production factor, information. Thus, as expected, the *substitution effect* between volume of information and labour is present;
2. the *complementarity effect* among contacts holds also in the case of n firms. The increase in the contacts between firm 2 and 3 decreases the number of contacts between firm 1 and 2. The magnitude of this effect for firm 1

depends on the efficiency of the contacts between firms 2 and 3. The higher the efficiency, the higher the decrease in the contacts between firm 1 and 2. This means in fact that firm 1 can obtain more information from firm 2, and thus can decrease the number of contact with firm 2, if firm 2 increases its contacts with firm 3;
3. the *costs effect* is still present in the model. The increase in the costs of use of these technologies for firm 1 leads to a decrease in the number of contacts between firms 1 and 2. The same holds for the equipment costs;
4. the *efficiency effect* still holds. The increase in the efficiency of contacts between firms 1 and 2 generates a decrease in the number of contacts between firm 1 and 2.

Our intent is to see whether these considerations may be confirmed also in other cases of connectivity. This is the subject matter of the next sections.

Partial connectivity. In this section we introduce a different case of connectivity, namely the case where firm 1 is connected to firms 2 and 3, but where the linkage between firms 2 and 3 is missing. The assumptions remain the same as in the previous case.

The *production function* has a Cobb-Douglas structure, containing the advantages firm 1 receives from being networked. These advantages are embodied in more information and an increase in the number of connected firms (that is, up to a certain number of subscribers, when the advantage of a marginal connection decreases, as shown in Chapter 4):

$$q_1 = \eta K_1^{\alpha_1} L_1^{\beta_1} N_1^{\delta_1} \qquad \text{with } \alpha_1 + \beta_1 + \delta_1 = 1$$
$$\text{and } \alpha_1, \beta_1, \delta_1 > 0$$

where K represents the capital stock of firm 1
   L represents the labour of firm 1
   N represents the network connectivity (or the volume of information) of firm 1

In this particular case, the *network connectivity* will be a function of the number of contacts between firms 1 and 2 generated by firm 1, represented by $I_{1,2}$, and generated by firm 2, represented by $I_{2,1}$. Moreover, the number of contacts will depend on the number of contacts between firm 1 and 3, represented by $I_{1,3}$ if generated by firm 1 and $I_{3,1}$ if generated by firm 3. The network connectivity function will be the following:

$$N_1 = \varepsilon I_{1,j} + \mu I_{j,1}$$

where $\varepsilon I_{1,j} = \varepsilon_{1,2} I_{1,2} + \varepsilon_{1,3} I_{1,3}$

and $\mu I_{j,1} = \mu_{2,1} I_{2,1} + \mu_{3,1} I_{3,1}$

The problem for firm 1 is thus to minimise costs under the assumption of a given production level. The procedure to obtain the result is similar to the previous cases (see Annex 1), and leads to the following result:

$$I_{1,j} = \frac{q_1^*}{\varepsilon \eta_1} \left(\frac{c_{L_1} \beta_1}{c_{K_1} \alpha_1}\right)^{\beta_1} \left(\frac{\delta_1}{\frac{\alpha_1 c_{I_1}}{\varepsilon c_{I_1}} + \alpha_1 c_{n_1}}\right)^{\alpha_1 + \beta_1} - \frac{\mu}{\varepsilon} I_{j,1}$$

Writing the results in terms of contacts of firm 1 only with firm 2, we obtain:

$$I_{1,2} = \frac{\varepsilon - \varepsilon_{1,3}}{\varepsilon_{1,2} - \varepsilon_{1,3}} I_{1,j}$$

and in terms of contacts of firm 1 only with firm 3, we obtain:

$$I_{1,3} = \frac{\varepsilon - \varepsilon_{1,2}}{\varepsilon_{1,3} - \varepsilon_{1,2}} I_{1,j}$$

Our mathematical results produce the following considerations:

1. the *substitution effect* is once more quite evident, as expected from the kind of production function used. The investments in the network increase when the costs of traditional production factors, capital and labour, decrease;
2. the *complementarity effect* among contacts also holds in this case. The increase in the contacts between firm 1 and 2 decreases the number of contacts between firm 1 and 3, and vice versa. The magnitude of this effect depends on the efficiency of the contacts between 1 and 3 for firm 1. The higher the efficiency, the higher the decrease in the contacts between 1 and 3. This is rather plausible, since the higher the efficiency of contacts with one firm, the lower the interest in receiving information from other firms;
3. the *cost effect* still plays a role. The increase in the costs of use of these technologies for firm 1 leads to a decrease in the number of contacts

between firm 1 and 2 or firm 1 and 3. The same holds for the equipment costs;
4. the *efficiency effect* is once more determined. The increase in the efficiency of contacts between firm 1 and 2 generates a decrease in the number of contacts between firm 1 and 3;
5. If the assumption that $\tau < \varepsilon$ holds, the number of contacts that firm 1 has to generate to achieve a certain volume of information is lower in this case than in the case of serial connectivity. It is easy to expect that the efficiency of a direct contact between two firms is greater than an indirect contact. For this reason we expect that a decrease in $I_{2,3}$ in the case of serial connectivity, or $I_{1,3}$ in the case of partial connectivity, means that $I_{1,2}$ increases less in the case of serial connectivity than in the case of partial connectivity.

Our attention is now focused on the case of full connectivity, where we expect our results to be confirmed, with some slight differences.

Full connectivity. This section contains the case of full connectivity, namely the case where the three firms are directly connected with one another. In this case the advantages of being networked should achieve their maximum. In fact, firm 1 gains from being networked to firm 2, to firm 3, and from the direct link between firms 2 and 3.

For the sake of comparison with the two previous cases, assumptions about the production function and cost function of networking remain the same. In particular, the *production function* is always a Cobb-Douglas production function, incorporating the advantages firm 1 receives from being networked. The *cost function of networking* is always dependent on two components: the use costs and the equipment costs, with the characteristics explained above. What is different from the previous cases is the *network connectivity function*. In this case firm 1 gains advantages from the connection with firm 2, with firm 3 and with the interconnection between firm 2 and 3. In this particular case, the volume of information will be a function of the number of contacts between 1 and 2, represented by $I_{1,2}$, of the number of contacts between 1 and 3, represented by $I_{1,3}$, and, finally, of the number of contacts between firm 2 and 3, represented by $I_{2,3}$:

$$N_1 = \varepsilon I_{1,j} + \mu I_{j,1} + \tau I_{2,3}$$

where $\varepsilon I_{1,j} = \varepsilon_{1,2} I_{1,2} + \varepsilon_{1,3} I_{1,3}$

and $\mu I_{j,1} = \mu_{2,1} I_{2,1} + \mu_{3,1} I_{3,1}$

The problem for firm 1 is thus to minimise costs under the assumption of the following given production level:

$$q_1 = \eta K_1^{\alpha_1} L_1^{\beta_1} N_1^{\delta_1} \qquad \text{with } \alpha_1 + \beta_1 + \delta_1 = 1$$
$$\text{and } \alpha_1, \beta_1, \delta_1 > 0$$

where K represents the capital stock of firm 1
L represents the labour of firm 1
N represents the network connectivity (or the volume of information) of firm 1

Solving the problem with the following Lagrangian function:

$$\mathcal{L}_1 = c_{K_1} K_1 + c_{L_1} L_1 + c_{I_{1,j}} I_{1,j} + c_{n_1} N_1 + \lambda_1 \left(q_1^* - \eta K_1^{\alpha_1} L_1^{\beta_1} N_1^{\delta_1} \right) + \lambda_2 \left(N_1 - \varepsilon I_{1,j} - \mu I_{j,1} - \tau I_{2,3} \right)$$

and minimising network connectivity costs, we find the following results:

$$I_{1,j} = \frac{q_1^*}{\varepsilon \eta} \left(\frac{c_{L_1} \beta_1}{c_{K_1} \alpha_1} \right)^{\beta_1} \left(\frac{\delta_1}{\frac{\alpha_1 c_{I_1}}{\varepsilon c_{I_1}} + \alpha_1 c_{n_1}} \right)^{\alpha_1 + \beta_1} - \frac{\mu}{\varepsilon} I_{j,1} - \frac{\tau}{\varepsilon} I_{2,3}$$

Writing the results in terms of contacts of firm 1 only with firm 2, we obtain:

$$I_{1,2} = \frac{\varepsilon - \varepsilon_{1,3}}{\varepsilon_{1,2} - \varepsilon_{1,3}} I_{1,j}$$

and in terms of contacts of firm 1 only with firm 3, we obtain:

$$I_{1,3} = \frac{\varepsilon - \varepsilon_{1,2}}{\varepsilon_{1,3} - \varepsilon_{1,2}} I_{1,j}$$

Some considerations can be drawn from these results:

1. the *substitution effect* is once more present. The costs of the traditional labour factor negatively influence the network connectivity level, which increases when the costs of production factors increase.
2. the *complementarity effect* among contacts also holds in this case. The increase in the contacts between firm 1 and 2 decreases the number of contacts between firm 1 and 3, and vice versa. The magnitude of this effect depends on the efficiency of the contacts for firm 1. The higher the efficiency, the greater the decrease in the contacts. This is rather plausible, since the higher the efficiency of contacts with one firm, the lower the interest in receiving information by other firms;
3. the *cost effect* still plays a role. The increase in the costs of use of these technologies for firm 1 leads to a decrease in the number of contacts between firm 1 and 2 or firm 1 and 3. The same holds for the equipment costs;
4. the *efficiency effect* is once more determined. The increase in the efficiency of contacts between firm 1 and 2 generates a decrease in the number of contacts between firm 1 and 3;
5. In this case the number of contacts firm 1 has to generate with other firms is lower than in the case of partial and full connectivity. Being connected with a greater number of firms allows firm 1 to achieve the same volume of information with lower costs.

The difference between the full connectivity case and the serial and partial connectivity cases is that if the number of contacts between firm 1 and 2 decreases, the number of contacts between firm 1 and 3 has to increase less in the full connectivity case than in the latter case, because of the presence of a number of contacts between 2 and 3. In other words, with of full connectivity, firm 1 receives the greatest advantages of being networked. This reflects the network externality concept, since the higher the number of firms connected, the greater the advantages a firm may get from being networked.

Before summarising our results and drawing some propositions to be tested empirically, some considerations are required concerning the case of a game among n asymmetric agents.

5.4.2. A sequential game with n firms: some considerations

The generalisation of the Von Stackelberg model, presented in Section 5.3.2., to the case of n firms is not so evident. In fact, contrary to the case of the two firm model, where the distinction between a leader and a follower is straightforward, in the case of n firms some hypotheses have to be included in the analysis and some considerations are needed.

First of all, the Von Stackelberg model itself loses most of its attractiveness and economic interest when applied to n firms. The logic underpinning the model is the conflicting positions of the two firms, and the equilibrium solution rests on the behaviour of the two asymmetric agents. An expansion of the model to the case of n firms does not add anything new to the analysis, when an assumption is made that one firm is a leader and n-1 firms are followers in a Cournot-type of competition. Thus, if we assume the existence of a leading firm and of many small firms depending on it (as in the case of the computer science market, with IBM dominating a market of many small price taking firms), the situation can easily be analysed as for the two-firm model, having one large firm in a dominating position, and n-1 small Cournot competing firms. This case would not add anything new to our analysis.

Another case would be to remove the assumption of n-1 Cournot competitors and make the assumptions of a chain of relationships, all characterised by leader/follower positions influencing the behaviour of the different firms. A methodology of analysis of this kind of market relation can be developed using network structure and network evolution models. This is certainly an interesting and stimulating area of research, but it moves away from our theoretical exercise and the purposes of our analysis. For this reason, we do not wish to enter this debate, and remain in the realms of more traditional literature of production theory and network externalities.

5.5. The case of telecommunications networks

The mathematical exercise developed in this chapter aims to produce some qualitative propositions which have to be proved empirically in the second part of this book. Before presenting these propositions, it is useful to make some reflections on the degree of realism of the results we achieved in relation to the case of telecommunications networks. In fact, quite interesting results are achieved from this perspective.

The first result is the "*substitution effect*" between the traditional labour factor and the level of connectivity. This result is perfectly realistic in the case of telecommunications networks. When the costs of labour increase, the firm is inclined to either find labour in areas where labour cost is not so high, or find a way of externalising some particular production functions, if labour involved in that particular function has increased its cost. For both these solutions, the existence of a telecommunications network is very helpful. In the case of finding new people in areas where labour is not so expensive, a telecommunications network is a useful tool. All forms of teleworking are an example of a temporary and remote labour force. As far as the externalisation of jobs or functions is concerned it is realistic to expect an increase in contacts via a telecommunications network with other co-operative firms.

The second result achieved, i.e. the *complementarity effect* among contacts, needs more reflection. An increase in the contacts on a telecommunications network between firm 1 and 2 decreases the number of contacts between firm 1 and 3 (in the case of full connectivity). In the case of a telecommunications network, this complementary effect can take place for two reasons. The first one is related to the technical capacity of a network. If this is fixed, an increase in the contacts between firm 1 and 2 may create congestion on the network and may influence the number of contacts (i.e. telephone calls) which firm 1 may have with other firms. A second reason may be related to the interest of firm 1 in getting into contact with other firms, when the information it receives from firm 2 increases. It can very well be the case that the information firm 1 receives from firm 2 substitutes other information firm 1 could get from other agents. However, this mechanism would need a deeper analysis of the kinds of means of contacts and of the actors involved, which is beyond the scope of this study and would require a particular analysis.

The results also show a *cost effect* involved, i.e. an increase in the costs of use of these technologies for firm 1 leads to a decrease between firm 1 and the other firms. If we apply this result to the telecommunications sector, the results of the previous mathematical exercise may easily be the case. An increase in the tariff of telephone calls may generate a disincentive to use the telephone. This definitely holds in the case of individuals[9]; it is less obvious and direct in the case of firms, where the use of the telephone is so vital that the demand is rather inelastic to changes in price. Even for firms, however, the "cost effect" can be valid for particular services, especially for very new services whose use is not yet so strategic in business relationships. Again, the generalisation of the results to all actors, all means and circumstances does permit a specific answer to the question whether or not a cost effect would take place in the telecommunications sector. Even in this case, this subject represents in itself an interesting area of study.

Another result from our mathematical exercise is the "*efficiency effect*"; an increase in the efficiency of contacts between two firms decreases the number of contacts these two firms have with other agents. This mechanism is plausible when related to a telecommunications network. The increase in the efficiency of a message received via a network may easily reduce the need to look for other information from other firms and agents, when the message obtained is comprehensive and useful enough.

The previous remarks allow us to say that, although some degree of uncertainty in some cases exists which requires a specific study in order to be solved, there is a substantial degree of realism in our results, when applied to the telecommunications technologies. The next task is to provide a clear account of the propositions derived, in order to test them empirically in the second part of this book.

5.6. Limits of the model

The present model is based on some assumptions, which inevitably limit the analysis. Before presenting the propositions deduced from our mathematical analysis to be tested at the empirical level (see Section 5.7), this section discusses the limiting assumptions on which our model rests.

First of all, in our model it is specifically assumed that the volume of information influences the production function, but with no clear distinction about the *quality* of these contacts. In the real world, of course, not only the quantity of information but also its quality is strategic for the effects information has on the production function, which may define the intensity of impact of the contact. In our model this aspect is only partially and indirectly treated via the parameters representing the efficiency of contacts in terms of the volume of information entering the production function (ε in the case of the two firms model; ε, μ and τ, in the model with three firms). A poor quality of information has a very limited impact on the volume of information. The extreme case is represented by an efficiency parameter equal to zero, showing a very low quality. If the efficiency parameter has a value equal to zero, the contact between firms has no influence on the volume of information introduced in the production function. However, our model does not directly refer to the quality of information exchanged.

Moreover, the same influence on the production function of firm 1 is given to the contact generated by firm 1 towards firm 2 and to the contact generated by firm 2 towards firm 1. It would be reasonable however to expect that the contact generated by firm 1 has greater influence on the production function than a contact generated by firm 2. In the former case, it may very well be the case that the information required by firm 1 is strategic information oriented towards the solution of some business constraints, while in the latter case, the probability that the information obtained by firm 1 from firm 2 via the network is strategic for firm 1 is lower than in the previous case. Our model takes this fact into account by attributing a different value to each efficiency parameter of the networking function. In the case of a networking function of firm 1, the efficiency parameter value is higher in the case of the contact generated by firm 1 towards firm 2, than in the case of the contact generated by firm 2 towards firm 1. In this way, the different weight of the contact is indirectly taken into consideration.

A third limit of our model is that there is no clear reference to the *characteristics of firms*. The analysis could take into consideration a typology of actors from whom the information could come, or to whom it could be sent, and attribute to each category of actors a different weight in the way information is exploited at the production level. It would be in fact reasonable to allot a different importance to a contact with a supplier than to one with a competitor or a customer.

Another limit is the absence in our model of the *technological characteristics* of the network on which the information is exchanged. In reality, one could introduce in the model a different efficiency, or cost of the contacts, depending on the network used. A digital network has, *ceteris paribus*, a greater efficiency in terms of volume of information exchanged, than an analogue network. The same can be said for different applications, which could be used on the network with different efficiency in the information transmission.

Moreover, our model does not take into account the *knowledge* necessary to use these technologies in an efficient way. As we will see in Chapter 9, the organisational capabilities and know-how to use these technologies are not really a public good, and vary among firms and regions, yet the existence of this know-how is actually of strategic importance in order to exploit production network externalities. This fact will be taken into account extensively in the empirical part of this study.

Last, but not least, as we have already mentioned before, another limit of the model is that the "substitution effect" among production factors is the result of the structure of the chosen production function.

These limits do not, however, decrease the value of our analysis, when the object of the exercise is taken into account. Our analysis has been made with the aim of defining some general propositions to be tested, based on a standard neoclassical approach. In the empirical part of the study, we will test these propositions, and extend the analysis to some more realistic cases, taking part of the previous limits into consideration in order to overcome them.

5.7. Results of the analysis: propositions to be tested

The aim of this section is to link the mathematical exercise developed above with the empirical analysis to be developed in the second part of this book. The mathematical analysis allows for the identification of some *propositions* which have to be tested empirically in order to prove the empirical validity of all that has been said at a theoretical and conceptual level.

The first result, common to all cases we presented, is the *substitution effect* between traditional production factors, capital and labour, and the expected volume of information: the volume of information increases when the costs of capital and/or labour increase. This result is influenced by the choice of the shape of our production function, as we have already discussed in a previous section.

A second result is the increase in the *efficiency of contacts* leading to a decrease in the number of contacts required by the firm.

A third result is the relationship between the *marginal productivity* of traditional production factors and the number of contacts between two firms.

The increase in the productivity of capital and/or labour leads to a decrease in the number of contacts required by the firm.

A *production network externality* effect emerges from our analysis, since a firm gains in terms of productivity even if the efforts to increase productivity are not made directly by the firm itself. The network externality effect is witnessed also by the fact that firm 1 decreases its number of contacts (i.e. its efforts to achieve a greater volume of information) via the network when the other connected firm increases its own efforts to communicate with firm 1.

A *cost effect* has emerged quite clearly, since in all cases there is a negative relation between the use and equipment costs of the network and the number of contacts (which is a measure of the intensity of use of these telecommunications technologies).

These results lead to the formulation of *specific propositions*, which will be tested empirically in the next part of this book. In particular, our mathematical model reaffirms once again what we suggested in the conceptual framework, namely:

1. *firms achieve greater levels of productivity if they are linked to the network. In more general terms, firms gain positive advantages for their production function because of their physical linkage to a network;*
2. *if a firm is in a leading position in the market, it exploits the advantages of its leading position in the same way as it exploits network externalities. This is shown by the fact that the increase in the efficiency of contacts generated by the follower for the leader generates a decrease in the number of contacts generated by the follower, and thus the number of contacts generated by the leader increases. In the case of a leader in the market, the number of contacts increases less than the number of contacts in the case of two symmetric firms;*
3. *the network externality advantages are greater, the higher the number of firms connected;*
4. *the advantages a firm gets from being networked are typical "network effects" and not "technology effects" (see Section 5.2.1. for the distinction between the two). In fact, although the results are a greater productivity of traditional production factors which could be associated with any technological innovation, in this particular case the increase in productivity is not always paid for by the firm itself. In all our cases, greater levels of productivity are achieved thanks to the expansion in the contacts whose costs are borne by other connected firms. This means that the marginal benefits achieved by the firm (greater productivity) do not reflect equal increases in marginal costs.*

These propositions may seem oversimplified in the light of our mathematical formulation. However, they seem to be sufficient for our purposes, in that they underline some strategic characteristics which have to be tested empirically in order to give validity to our conceptual framework. More sophisticated results would in this case be of no use because of the data limitations constraining our empirical analysis. The most serious limitation of our database is that it is not a time series database, thus precluding any dynamic analysis.

For this reason, we keep the present mathematical exercise in its present form, while being aware that at least two possibilities remain open for further research:

a) a dynamic analysis of the model;
b) a test of more specific propositions with a more powerful database.

Despite all the limitations of our mathematical exercise, the purpose for which it has been developed has been fulfilled. In fact, the definition of these propositions arrived at allows us to test our theoretical "economic and spatial symbiosis" model empirically on the basis of a rigorous economic framework.

5.8. Conclusions

The attempts made in this chapter were geared to finding a link between our conceptual framework and our empirical work, by using traditional micro-economic tools. The use of these instruments provides a firm base for our conceptual framework and serves as a bridge to link this framework with the empirical analysis which follows in the next chapters. In fact, by introducing the concept of network externalities into the traditional micro-economic production theory at least two objectives are achieved. The first, already mentioned, is to give to our conceptual framework strong support from the classical and well-known micro-economic theory of the firm. The second, as important as the first, is to arrive at the formulation of some qualitative testable propositions. The extent to which these propositions can be verified or refuted gives a measure of the empirical validity of our framework.

Thus, the mathematical exercise was not meant to generate an econometrically testable model. On the contrary, it was oriented towards the qualitative results obtained by a simple analysis of the relationships among different network-oriented variables. From this perspective, the exercise has been rather successful and some propositions have been derived.

In the next chapter we will prepare the ground for the empirical analysis. We will link these propositions to the research issues we raised at the beginning of the book (see Chapter 2). After this link has been specified and some testable hypotheses formulated, we then have a solid background for our empirical analysis, which will be presented in Chapters 7, 8 and 9.

Notes

1. See Blankart and Knieps (1992) for a similar distinction in the case of consumption network externalities.
2. A similar analysis has been carried out by Rietvelt, Rossera and van Nierop (1993) in the case of substitution/complementarity effects among communication modes.
3. The CES production function has the form:

$$y = \left[a_1 x_1^\rho + a_2 x_2^\rho \right]^{(1/\rho)}$$

It is easy to verify that if $\rho = 1$, the CES function takes the form of a linear function. If $\rho = 0$, CES takes the form of a Cobb-Douglas function, and if $\rho = -$ infinitive, CES takes the form of a Leontief function. For a mathematical verification, see Varian, 1992; for an empirical application see Rietveld, 1987.
4. In this section the output has been labelled "q" as is usually the case in the production functions modelling, but it is the same output expressed as "Y" in the previous sections.
5. Given the fact that in our case both prices and quantities are fixed, these levels of input factors guarantee not only the minimisation of costs, but also the maximisation of profits.
6. For the mathematical solution of the equation systems see Annex 1 at the end of the book.
7. The second firm in the market is Fujitsu, with a turnover of nearly 4,000 million US $ in 1991 (Source: Mondo Economico, 24 July 1993).
8. For the mathematical solution of the equation systems see Annex 1 at the end of the book.
9. Research on demand elasticity of telecommunication services is taking place at the Economics Department of the Politecnico of Milan, sponsored by the Italian telecommunication public provider, Sip. Although the results are in a preliminary form, the demand price elasticity of telephone calls seems to be rather inelastic, especially in the case of international telephone calls (Colombino, 1993).

Part B
EMPIRICAL INVESTIGATIONS

6 An empirical framework for network externality measurement: The case of Italy

6.1. Introduction

The first part of our study (Chapters 1-5) dealt with the concept of network externality and its possible impact on industrial and regional performance. Based on a broad literature survey, we have highlighted the still unexplored areas of research related to the network externality concept, viz. the relationship between network externality and the production function, at both a micro- (firm) and macro- (regional) level. In this context, a conceptual framework has been formulated, the "economic and spatial symbiosis" framework, explaining from a conceptual point of view the possible linkage between network externalities and the performance of firms and regions. This framework has been supported by a formal mathematical model, highlighting theoretically the role network externalities play in the production function and its subsequent results.

The second part of our study addresses the issue of the empirical validity of the above mentioned "economic and spatial symbiosis" framework. For this purpose, the aim of this part of the study is to test empirically the hypotheses on which our analytical framework rests. In particular, these assumptions will be derived from our conceptual framework and from its mathematical representation. As stated in Chapter 5, the aim of the model was to achieve a systematic and logical representation of our conceptual framework, by means of a mathematical exercise. The mathematical model is then mainly used to define a set of qualitative propositions related to the "economic and spatial symbiosis" framework, and gives rise to some basic propositions that have to be verified (refuted), in order to test empirically the validity (the non-validity) of our model.

In particular, as discussed in Chapter 2, our research questions, which have to be investigated empirically, are whether:

a) an important reason for telecommunications adoption rests on the number of already existing subscribers;
b) there is a (direct and indirect) link between the degree of connectivity to the network of a firm (and a region) and its performance;
c) there are some micro- and macro-conditions facilitating or hampering the exploitation of network externalities.

The aim of the present chapter is to highlight the results obtained from our conceptual framework and from our mathematical model and to "translate" them into testable hypotheses which will be investigated in Chapters 7, 8 and 9. Thus, the present chapter provides the *logical bridge* between what has been theoretically and mathematically said in the first part of the book, and the empirical analysis to be presented in the following chapters.

Our empirical analysis concerns the impact that advanced telecommunications technologies may have on industrial and regional performance. The evaluation of telecommunications projects at a regional level is therefore the most appropriate empirical investigation to be undertaken. In the light of this, the empirical analysis will be devoted to the evaluation of the Special Telecommunications Action for Regional Development (STAR) programme, which has been run with the aim of developing advanced tele-communications technologies in less developed regions of the European Community (EC)[1]. In particular, within the STAR programme, our analysis concentrates on the case of Southern Italy, where a sample of small- and medium-sized enterprises have been interviewed. In order to include the regional dimension properly, we also developed the same empirical analysis for firms located in the North of Italy, an area with a much higher economic development. The contrasting economic situation allows the comparison of network externality effects in different economic environments; these differences could not have been measured to the same degree if the analysis had been carried out entirely in other less favoured regions of the European Community.

The reason for the choice of small- and medium-sized enterprises is explained by the fact that within small- and medium-sized firms the effects of new telecommunications adoption on the competitiveness of firms (see Chapter 4) are expected to be more evident. In large firms the identification of the direct impact of these technologies on the competitiveness of firms is rather difficult to measure. Large firms have in fact many more channels through which they may increase their competitiveness, than SMEs, and our exercise would become more difficult than it is already.

The rest of the present chapter is devoted to the presentation of the empirical analysis. Since the main subject of the analysis is the evaluation of the STAR programme, Section 6.2 contains a brief description of the programme, how it has been developed, run and managed, in order to understand the extent of EC intervention at both the organisational and financial level. Section 6.3 contains an in-depth analysis of the testable hypotheses derived from our conceptual and mathematical framework, which will be investigated empirically in Chapters 7, 8 and 9. Section 6.3 also gives an outline of a telecommunications impact analysis on the industrial and regional performance. This *"causal path analysis"* captures the relationship between telecommunications adoption and (industrial and) regional performance. Section 6.4 contains a brief presentation of the database, with particular emphasis on the reasons for the choice of this database. Some concluding remarks are presented in Section 6.5.

6.2. Telecommunications and regional development: the STAR programme

6.2.1. General features of the programme

The increasing awareness of the importance of telecommunications infrastructures and services for encouraging economic development has been a prominent feature of the 1980s, as is witnessed by an increasing body of economic literature underlining the importance of telecommunications technologies for economic growth at both a theoretical and an empirical level. This increasing awareness has also stimulated policy interventions in this direction, by interpreting infrastructure in general, and telecommunications technologies in particular, as a power source of economic growth.

The Special Telecommunications Action for Regional Development (STAR) is a European Community programme launched in 1987, reflecting the general positive attitude towards telecommunications infrastructures as the key elements in the pursuit of regional development. Under the EC auspices, the STAR programme aims to encourage economic development in the less favoured regions of the Community by means of easier and quicker access to advanced telecommunications technologies. The programme was launched in all the less favoured regions of the Community, the so-called "Objective 1" regions, namely part of Italy, of Spain, of Great Britain, of France, and the whole of the Republic of Ireland, Greece and Portugal. Figure 6.1 presents a map of European regions, where the hatched areas represent the "Objective 1" regions[2].

The backwardness of the less-favoured regions in telecommunications infrastructure and services was the second reason for this direct intervention by the Community. In 1987, when the STAR programme was launched, the distribution and quantity of the telecommunications infrastructures in those regions was rather poor.

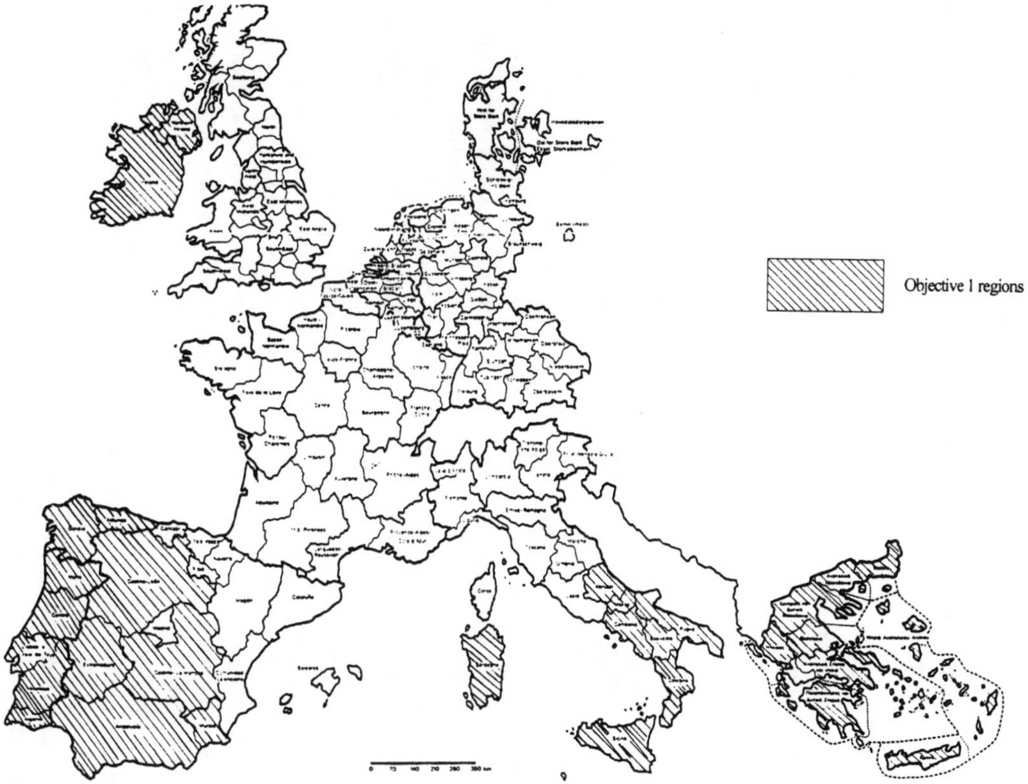

Figure 6.1. Map of 'Objective 1' regions

The telephone network density was extremely low in comparison with core regions, and the situation was even worse for advanced telecommunications networks and services.

The intervention of the Community aimed at diminishing the disparity in terms of network accessibility between the advanced and less-favoured regions, by installing basic infrastructures (optical fibre networks, satellite networks, integrated service digital networks - ISDN -, advanced networks) for advanced telecommunications services and by encouraging the supply and demand for such services. Thus, the STAR programme is a typical supply driven innovation process, aiming at:

a) *developing a technological supply* in areas where the market rules would never have encouraged public operators to make sufficient financial efforts. In fact, the lack of an explicit demand for advanced telecommunications services in these regions does not in fact justify the financial effort required to produce these advanced technologies. However, the project of implementing advanced telecommunications networks has been developed with the full involvement of the public operators for different reasons. First of all, the institutional monopolistic position of most of the European public operators (apart from the British case) would otherwise have meant that it was not possible to realise a full implementation of the programme. Secondly, the "additionality" principle governing the European Structural Funds would otherwise not have been fulfilled: the Community in fact expects, as a basic principle for its local policies and interventions, the same financial effort from the National bodies in areas where otherwise this kind of investment would never take place;

b) *promoting advanced and innovative telecommunications services*, with the help of pilot services in specific micro-areas;

c) *supporting and stimulating a demand for these services*, in order to create a real market at the end of the five-year programme. For this reason demonstration and promotion centres have been developed in order to facilitate the linkage between technological potentialities of the supply and the real needs of the demand. In other words, these centres play the role of intermediaries in the market, with the aim of stimulating a sustainable innovation process. While the promotion of advanced services is based on a sample of firms, this phase is directed to all potential business users.

In concrete terms, the STAR programme has developed:

- a broadband network;
- experimentation of a the Integrated Service Digital Network (ISDN);
- specialised networks (packet switching network, digital switching network, digital circuits, fib-re optic networks);

- satellite network to support the development of advanced services in specific demonstration centres;
- other more specific interventions in some areas;
- five specialised centres for assistance and training in the use of new telecommunications services (such as video conference; data banks; on-line CAD/CAM technologies, etc.).

A large group of SMEs in the South of Italy was chosen through an ex-ante evaluation of the potentially most innovative firms in South of Italy, based on indicators of their technological and innovative development pattern and their entrepreneurial capabilities. To these firms an on-line connection to the advanced telecommunications services available at the demonstration centres, and sometimes access to the advanced networks (when ready), was given. All services and on-line connections were free of charge. Thus, these firms could use CAD/CAM systems on their (on-line) personal computer, could have access to data banks (different for the different industries) and could exploit the electronic mail service.

In order to develop all networks and services mentioned above, the programme has been organised in two large projects aiming at:

- developing advanced telecommunications networks (fibre optic networks, Integrated Service Digital Network, digital telephone networks) for the development of advanced services (the so-called 4.1 measures)[3]. Unfortunately, these networks still have a very low level of connection with networks in the North;
- developing promotion and demonstration centres (the so-called 4.2 measures).

The structure of the programme is presented in Figure 6.2. The programme was run by DG. XVI (Regional Policy) of the EC with a small peripheral role played by DG. XIII (Science and Technology). 50% of the available funds were spent by DG. XVI for both the 4.1 measures (infrastructure provision) and 4.2 measures (service promotion).

A National STAR Steering Committee has been created for monitoring both 4.1 and 4.2 measures. In the case of service provision, the National Committee was in charge of the decision on the kinds of projects to be developed under the 4.2 measures, which had to be additionally financed with national public funds (for the remaining 50%). In the case of the 4.1 measures, the national public operator has of course been playing a strategic role, by being obliged to finance the other 50% of the 4.1 measures. The important role played by national bodies in all countries has led to a different configuration of the STAR programme itself in the different Countries.

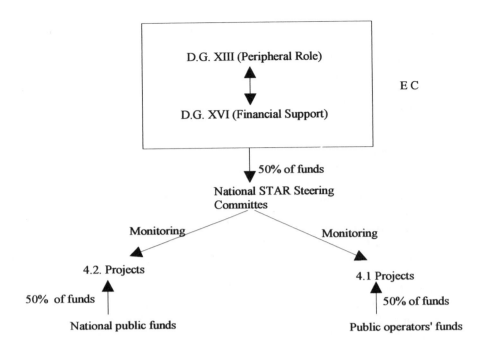

Figure 6.2. Structure of the STAR programme

A simple example of this difference is provided by comparing the Italian case with the Spanish. The logic underpinning the STAR interventions in these two countries is rather a contrasting one. In Italy, national bodies opted for a *"top-down"* intervention, i.e. the decision about the type of network or service was made at a national level, and then applied in the different areas. In Spain, in contrast, the policy chosen for the promotion of the STAR technologies was quite the opposite, i.e. a *"bottom-up"* approach, where the decisions about which networks and services had to be offered were made at the local level by the local responsible bodies. The resulting experience in terms of networks and services offered in Italy and Spain has been rather different. In Italy, the "top-down" approach has brought about the development of similar new networks and services in all regions of the Mezzogiorno, but with no particular interest in local needs or local requirements as expressed by local users, but with the possibility of connecting all Southern Italy with well advanced and modern infrastructures. In Spain, on the contrary, the "bottom-up" policy has led to the creation of networks and services customised to meet all specific local requirements and needs, but with no possibility for long distance communications.

The allocation of resources to different projects reflects the different adoption philosophies and policies, thus following different patterns in the different countries. Table 6.1 shows the amount of financial resources allocated by countries for the different kinds of interventions.

6.2.2. The choice of the sample

The existence of the STAR programme has been of great value for our empirical analysis. In fact, our analysis aims at capturing what we called the *"network effect"*, i.e. the effects that the access to an advanced telecommunications network have on the competitiveness of firms. Firms in the sample of Southern Italy are all STAR-technology potential adopters: the sample is biased with respect to the entire population of Southern enterprises but, as desired, is representative of the potentially interested firms.

The reasons for the choice of the STAR programme as the basis for the empirical analysis in the South are manifold:

a) first of all, these telecommunications technologies are still in the *first phases of adoption*. This facilitates the definition of the most important stimulus of adoption. Our first hypothesis can, therefore, find without any doubt the right seedbed conditions to be investigated empirically;
b) the second reason for this choice is that this programme is devoted to *the development of advanced telecommunications technologies in the most backward regions of the Community*, with the aim of stimulating regional and local sustainable development. Hence, the evaluation of the real effects of telecommunications on this regional development is more evident and more realistic to perform than in more developed regions, where the marginal benefits of corporate and regional competitiveness are more limited;
c) the third reason is related to the fact that these telecommunications technologies are developed free of charge, and in this way *prices are not taken into account*. Price elements could thus be a misleading variable in our analysis; clearly, via the tariff structure, some of the network externality advantages may in some sense be internalised in the price system. However, in our opinion the tariff structure in telecommunications networks is unable to internalise all network externality effects. In fact, the advantages a firm obtains from being networked may be only partially paid for via the tariff structure. "Collect calls" systems, where the caller does not pay for the service used, are an example of the fact that the costs of the use of an additional unit of service does not always reflect the marginal benefits. Moreover, a user (i.e. a firm) may gain from an exchange of information, when the call is paid by another user (firm). In this case, again, marginal benefits do not reflect marginal costs;

Table 6.1
Total expenditure on STAR projects - 1993

DESCRIPTION	FRANCE	GREECE	IRELAND	ITALY	SPAIN	PORTUGAL	U.K.	EEC TOTAL
Optical Fiber	19%	3%	7%	57%	-	-	17%	21%
Satellite	-	1%	5%	2%	-	-	-	1%
Microwave	-	-	-	-	3%	-	-	1%
Digitalisation	6%	68%	25%	9%	54%	52%	66%	37%
Data Networks	21%	5%	8%	9%	21%	5%	-	11%
Mobile	10%	-	34%	-	-	9%	-	4%
Test Labs.	1%	1%	1%	1%	3%	-	-	1%
Studies	0%	1%	1%	0%	1%	-	-	1%
TOTAL 4.1	57%	77%	79%	78%	82%	67%	83%	76%
Studies	4%	2%	3%	0%	2%	1%	1%	1%
Marketing	6%	1%	1%	7%	3%	5%	0%	4%
Applications	33%	20%	17%	15%	14%	28%	15%	18%
TOTAL 4.2	44%	23%	21%	22%	18%	34%	17%	24%
TOTAL (in million ECU)	51,8	188,6	96,9	534,5	378,3	290,1	22,7	1563

Source: European Community

d) a final reason for the choice of this empirical analysis lies in the fact that these networks are *public networks*, and thus by definition they embody the concept of network externalities, since everybody has the right to join the network and to gain from network externality effects (see Chapter 2). No economic or institutional barriers protect the achievement and exploitation of network externalities, as in the case of exclusive networks. Moreover, private networks are excluded from the analysis, since their being a private good inhibits the opportunity of exploiting network externalities.

For these reasons, the STAR programme provided an interesting area of analysis for less developed regions. However, for our purposes of measuring network externality effects in different economic environments, a further analysis developed in a rather contrasting economic *milieu* was necessary. As a case of a developed region we have chosen the *North of Italy*, which contrasts sharply with the South. In fact, the diffusion process of new technologies is more advanced; firms are stimulated by a more developed economic environment, and, last but not least, these technologies, although public technologies, are commercial technologies. These contrasting characteristics undoubtedly give much more weight to a comparative study, since they allow:

a) the measurement of network externalities as mechanisms for technological adoption in different stages of the diffusion process. In the case of Northern Italy, the development of these technologies is much more advanced than in Southern Italy. Thus, the analysis in both geographical areas will allow us to test the importance of network externalities as reasons for adoption in different stages of diffusion;
b) the measurement of the capacities of exploiting network externalities in different economic environments. In this way we are able to understand under which macro-conditions firms are more able to exploit network externalities.

Some interesting results stem from the analysis. Especially, what emerges is the different behaviour of the two macro-areas in the choice of the most important reasons explaining the adoption process of telecommunications technologies (Chapter 7). Moreover, as Chapter 8 will highlight, a strong inter-regional difference also exists in the way these technologies are used and their effects in the production function.

6.3. Methodological framework for the empirical analysis

This section serves to provide insight into the way the empirical validity of the "economic and spatial symbiosis" framework will be verified. To this end some testable hypotheses will be presented, derived from our "economic and spatial symbiosis" framework sketched in Chapter 4 and from our mathematical model. Clearly, data limitations - concerning the adoption of these technologies over time, and the spatio-temporal dimension of the adoption process of telecommunications technologies - preclude a statistical/ econometric empirical testing of the "economic and spatial symbiosis" framework. The hypotheses which will be deduced in this section relate directly to the research issues presented in Chapter 2. In this way we highlight the link between the research issues and the testable hypotheses for our empirical approach.

The empirical analysis was carried out through in-depth interviews with corporate users, with a structured questionnaire. The logical structure of the questionnaire[4] is presented in Figure 6.3, where the underlying framework of the empirical analysis is summarised.

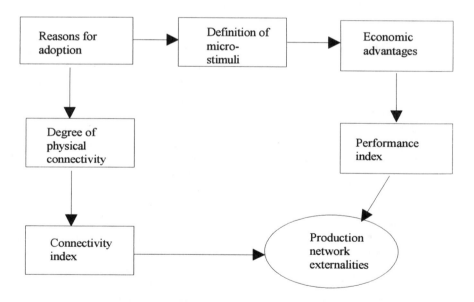

Figure 6.3. Structure of telecommunications technologies' impact analysis of economic and spatial symbiosis

The first research issue was to see whether the micro-stimulus generating the adoption process of telecommunications may be associated with network externalities. In other words,

"whether one of the main reasons for joining the telecommunications network is the number of already existing subscribers"

For this purpose, the first part of the questionnaire deals with the *reasons and bottlenecks* concerned with the adoption of these technologies. Our model suggests that in the first phases of a diffusion process:

1) the main reason for non-adoption is the low number of subscribers already linked to the network;
2) the main reason for the adoption of telecommunications networks and services is not the number of already existing subscribers;
3) a positive relation exists between a larger number of subscribers and the decision to adopt by potential users;
4) the major problem still existing in the use of new telecommunications networks and services is the low number of subscribers;
5) the main reasons for non-increasing intensity of use is the low number of adopters.

These hypotheses are expected to hold in *the first phases of diffusion*. With a behavioural analysis on telecommunications users it is possible to verify whether our hypotheses, stemming from our "economic and spatial symbiosis" framework, are verified or refuted. In particular, we run a statistical analysis on the reasons for the choice of adoption or of non-adoption of telecommunications technologies, the degree of satisfaction of these technologies by users and the problems still existing in the use of these technologies. This is the subject matter of Chapter 7.

The second part of the questionnaire deals with the *economic advantages* obtained by these firms from the use of these advanced telecommunications technologies and services in terms of more economic relationships with other firms, and in terms of better economic performance. This part allows for the definition of the economic advantages for adopters, in both quantitative and qualitative terms. This part of the questionnaire highlights the advantages obtained by joining the network, as well as the judgements on the importance of telecommunications technologies for business development and corporate competitiveness. In this way, a synthetic performance index may be constructed and - if a positive relationship is envisaged between the intensity of use of these technologies and the economic advantages of a firm - these network externalities are supposed to have an impact on the dynamics of the economic performance. This empirical approach allows us to reply to our second crucial research question, namely:

"whether network externalities have an impact on the performance of firms".

In particular, from our conceptual and mathematical framework we deduce the following testable hypothesis:

6) a positive relationship exists between the intense use of multimodal connections (e.g. telephone, e-mail, fax) and the firm's performance; on the other hand, as we have already claimed theoretically, we expect that there is no direct relation between the rare use of multimodal connections and the firm's performance.

Another research question mentioned in Chapter 2 deals with the spatial dimension of the network externalities effects, namely:

"whether network externalities have an impact on the performance of regions".

At the regional level, the testable hypothesis deduced from our framework is the following:

7) a positive relationship exists between the use of multimodal connections (e.g. telephone, e-mail) and the regional performance. The regional aspect may well be captured as the result of a multiplicative effect of the number of firms located in the area which are able to exploit production network externalities. The sum of these positive effects will inevitably be reflected in the regional performance.

Finally, our analysis is developed in different regions within the same country, as well as in different sectors (in traditional manufacturing sectors and in more dynamic sectors) in order to highlight the best micro- and macro-conditions for the exploitation of the network externalities effects. Our third research issue is to define:

"under which micro- and macro-conditions network externalities play a role in the performance of firms and regions".

It would be too simple, and thus misleading, to try to identify a linear relationship between the availability of telecommunications infrastructure and corporate and regional development. A "causal path" analysis in the diffusion patterns of these technologies is, however, of crucial importance and this has been taken into consideration in the empirical analysis of our study.

Diffusion patterns of innovation are always the result of an interaction between supply and demand elements, whose better or worse interaction explains the rhythm of innovation development. The process may start from either the supply or the demand side, but the crucial element distinguishing the development patterns is what exactly is the interaction between the two parts involved.

In our empirical analysis within Southern Italy, the STAR programme is a typical supply driven innovation process. An "injection" of financial resources has dramatically increased the availability of the network and service access in Southern Italy in the last five years with the aim of increasing economic development. Table 6.1 showed that more than 534 million ECU have been invested in the STAR programme in Italy, with a clear preference for 4.1 measures: 78.4% of available funds have been devoted to the implementation of advanced networks. These funds have been used to implement a digital broadband network, an advanced widespread network linking different local area networks (LAN), and a packet switching network, called ITAPAC and already available in the rest of the country. As far as 4.2 measures - regarding the implementation of promotion and demonstration centres - are concerned, 21.6% of funds were allocated, allowing the development of five centres in the South of Italy, where demonstration activities for the most advanced services take place - i.e. video conference, sophisticated data banks and electronic mail. Some specific services, such as electronic mail and on-line data banks, have been offered free of charge to a number of potentially more dynamic SMEs, i.e. to those SMEs already having a personal computer in their organisation.

From the supply side this has meant an increase in the opportunity to connect more firms to advanced and sophisticated networks. This opportunity is translated into real connectivity only if the importance of these technologies for local business is clear to potential users. The importance of these technologies for business is dependent on two elements: a) the number of already connected business-related firms and b) the evidence that the new communication means provide more advantages than the old ways of communication. This critical phase needs strong support from the supply side. Its role is not only related to the implementation of physical networks, and thus physical access to new networks ultimately achieving a critical mass, but also to the promotion and demonstration to users that these new services and networks are vital for corporate competitiveness and for the achievement of competitive advantages. Cross-learning processes between demand and supply are seen as a crucial element determining the rate of diffusion of these technologies (Camagni and Capello, 1991; Capello and Williams, 1992).

In the logic of interaction between demand and supply we can also add spatial aspects. On the supply side, an increase in the availability of networks and thus in the possible connectivity of firms generates an increase in the locational advantage for firms in that area. In other words, these networks and services become a locational factor in themselves and may stimulate the location of new firms. The presence of a greater number of firms has an impact on the aggregate performance of the region as a whole. On the demand side, if the possible connectivity is translated into real connectivity, regions gain from the provision of scarce resources (for example, in backward regions, strategic information or know-how), which leads to an increase in the competitiveness of local firms. The sum of all these positive effects leads to a better regional performance.

In sum, it is clear that the relationship between connectivity and corporate and regional performance is far from being a linear and direct relationship. Many other elements play a role in the definition of this linkage. The "economic and spatial symbiosis" framework works under the assumption that all these elements play their role in the diffusion pattern. If just one of them is not working, the "economic and spatial symbiosis" framework cannot be demonstrated empirically.

From all that has been said above, we can deduce from our framework the following testable hypotheses at a micro level. Production network externalities are exploited if:

8) a critical mass of adopters is achieved;
9) an innovative use of these technologies is achieved, i.e. a use which does not only lead to an increase in the efficiency of the firm, but which is able to lead the firm towards better competitive position, through the achievement of new market niches, of product and process innovation;
10) support from the supply side is achieved in the first phases of diffusion;
11) entrepreneurship is present in the managerial staff of the firm;
12) a clear importance of these technologies for business purposes is envisaged by users.

At the macro-level, we expect production network externalities to be exploited by more advanced regions, since the micro-conditions expressed before are undoubtedly much more likely to be present in advanced regions.

Our empirical analysis, applied to different industries and regions, will also be able to identify whether all these hypotheses may be regarded as verified (or refuted) in particular industries and regions. Hence, Chapter 7 is devoted to the presentation of the results obtained through the behavioural analysis, which is useful to test the reasons for adoption (hypotheses 1-5). Chapter 8 is then devoted to test hypotheses 6 and 7 empirically, while Chapter 9 is devoted to test hypotheses 8-12.

6.4. The database

In this part of the chapter we will present the database on which we develop our empirical analysis (Chapters 7, 8 and 9). As mentioned before, the database is represented by primary data, collected via in-depth interviews with users of telecommunications technologies in two different macro-areas[5] in Italy: the North of Italy, in particular the highly industrialised Milan area, and the South of Italy, a typical area of economic underdevelopment.

The aim of the in-depth interviews was to capture the "*network effect*", i.e. the increase of access to telecommunications technologies and the degree of competitiveness which firms have acquired via the exploitation of these new technologies. In the case of the sample in the South, we have chosen to evaluate the effects of the STAR programme. An in-depth interviews methodology was chosen *inter alia* because in the case of the STAR programme it is difficult to separate the general effects of technological investments in new telecommunications networks and services from the particular effects of STAR specific investments. One way of dealing with this distinction is to make a direct inquiry based on a structured questionnaire among firms which are already using STAR's telecommunications technologies. Once this methodology had been chosen for Southern Italy, the same one had to be applied of course for Northern Italy.

The total number of interviews was 70, equally divided between Northern and Southern Italy (35 in each macro-area); they were carried out between September 1992 and February 1993 and were equally distributed between the manufacturing and the service sector (Table 6.2). In terms of intra-sectoral distribution, manufacturing firms belong to a varied range of industries, both traditional (food, construction), and modern (engineering, including electronics). In contrast, service firms have a high presence of business services (often computing and telecommunications) and a small minority of specialised trade services.

The choice to conduct an analysis in different sectors of the economy is necessary in order to deal with the research issue of whether specific sectoral conditions exist which influence the exploitation of network externalities. A difference is expected between service and manufacturing firms in the way they use telecommunications and, consequently, in the way they exploit network externalities.

Firms are, as required, small and medium-sized[6]: in 1991 the maximum employment per firm was 78 (with one exception of 160) in the South and 204 in the North; maximum turnover is 106 billion Italian lire in the South (approximately 58 million ECU) and 48 billion lire (26 million ECU) in the North.

Table 6.2
Sample structure

	Turnover	1991		
	0 - 500*	501 - 5000*	5001 - 10000*	10000 - 100000*
SOUTH	7	18	4	6
NORTH	0	13	6	15
ENTIRE SAMPLE	7	31	10	21

* Million Italian Lire - Current Prices

	Employment	1991		
	1 - 10*	11 - 20*	21 - 50*	51 - 200*
SOUTH	14	9	4	8
NORTH	3	6	14	11
ENTIRE SAMPLE	17	15	18	19

*Number of jobs

	Industry	Services
SOUTH	62.9%	37.1%
NORTH	42.9%	57.1%
ENTIRE SAMPLE	52.9%	47.1%

The distribution by turnover shows that the sample in the South is predominantly constituted by firms achieving a maximum turnover of 5 billion lire; on the contrary, in the North the distribution is in favour of firms having a minimum turnover of 5 billion lire (see Table 6.2). This difference appears also in the case of employees. In the South most firms have a maximum of 20 employees, while in the North most firms have more than 20 employees. This difference mirrors the discrepancies in the economic environment of the two macro-areas: a larger number of small firms in the South, compared to a larger number of medium-sized firms in the North. The different composition of the two samples of firms are taken into consideration when dealing with empirical analyses. In particular, some statistical tests are carried out in order to be sure that the different size and sectoral composition in the two samples does not affect our empirical results[7].

The reason for the choice of SMEs is that, as already explained before, we expect to find a clearer relation between new telecommunications technologies and industrial performance within SMEs than within large firms.

The questions asked to the sample of firms cover a high range of issues, namely the degree of satisfaction among users of the new services, the intensity of use, the degree of satisfaction with the marketing policies, the interest these services have raised among users, and the relationship between the introduction of these technologies and their business development. The whole questionnaire was divided into three different parts. The first part serves the purpose of testing the first group of hypotheses related to the first research issue; the questions in fact deal with:

a) the most important reasons for the adoption of telecommunications technologies;
b) the most important reasons for not having adopted telecommunications technologies;
c) the main conditions for future adoption of telecommunications technologies;
d) reasons for dissatisfaction with the telecommunications technologies in use.

The second group of questions is related to the second research issue and it focuses mainly on:

a) level of turnover;
b) number of employees;
c) telecommunications infrastructure and services used;
d) the intensity of use of these technologies;
e) number and kinds of innovation achieved via the introduction of telecommunications technologies;
f) number and kind of functions having faced organisational changes in the last five years;
g) variation in the number of suppliers and customers in the last five years.

Finally, another group of questions is related to the third group of hypotheses and concerns:

a) the degree of importance of demonstration and promotion centres in the decision-making process to adopt the new technologies;
b) the degree of importance of these technologies for the business.

To test the third research issue, other questions included in the questionnaire were also useful, namely the number of functions which had been subject to organisational changes and the amount of innovation achieved.

This broad range of questions was sufficient to create a comprehensive database on network behaviour on which we could build our empirical work.

6.5. Conclusions

This chapter has highlighted the link between our conceptual framework presented in Chapter 4, our mathematical model (Chapter 5) and the empirical part of the study which will be the subject matter of the rest of the book.

This link is represented by the identification of some *key testable hypotheses* derived from the conceptual and mathematical frameworks and which should be able to answer the key research questions of this study. The extent to which these hypotheses are verified (or refuted) is a measure of the empirical validity of our "economic and spatial symbiosis" framework.

In this chapter the database was also presented and especially the reasons for the choice of that particular database were underlined. All this facilitates the transition to the following empirical part of this work and the appreciation of the results stemming from it.

Notes

1. An international evaluation on all aspects of the STAR programme has been commissioned by the EC to Ewbank Preece, a British research centre, in collaboration with an Irish research centre (NEXUS) and the Bocconi University in Milan. The author of this book has participated directly in this general evaluation, as a team member. National evaluations have been separately commissioned by national goverments. For Italy, the national evaluation was carried out by CENSIS (Rome).
2. In the case of France, Objective 1 regions for the STAR programme were the islands of Guadeloupe and Martinique. They do not appear in the European map.
3. The EC has divided the intervention programme into two large projects. The first deals with the implementation of networks, and all interventions dealing with this project are called the "4.1 measures". The second deals with the implementation of demonstration and promotion centres for advanced services, and all interventions regarding this project are called "4.2 measures".

4. For a detailed description of the questions raised, see Annex 2 at the end of the book.
5. By "macro-areas" we mean several regions analysed together. In the case of the South of Italy, in fact, we analysed three Southern regions, namely Sicily, Abruzzi and Apulia. The results of the empirical analysis for these three regions are presented together under the heading "South of Italy". In the case of the North of Italy, instead, the empirical analysis was run only in one region, i.e. the Lombardy region, and the results are presented under the heading "North of Italy".
6. The STAR programme was set up only for SMEs. In this study our sample contains only small- and medium-sized firms: they have no more than 200 employees and a turnover of no more than 100 billion Italian lire.
7. See Chapters 7 and 8 for details on the statistical tests used.

7 Industrial and regional variations in consumption network externalities

7.1. Introduction

In the previous chapter we specified the testable hypotheses derived from our conceptual and mathematical framework. In this chapter, and in the next two (Chapters 8 and 9), we will present the results of our empirical analysis, in order to test whether our "economic and spatial symbiosis" framework is empirically valid. In particular, the present chapter draws attention to the first basic research issue to be investigated, i.e. whether *network externalities are one of the main reasons for entering the network*. Especially in the first phases of adoption, one of the most important incentives to adopt interrelated technologies is the number of already existing subscribers. As the traditional theory on network externalities underlines, the user value of a network is dependent on subscriber base.

As described in our "economic and spatial symbiosis" framework, one of the most important reasons for a firm to join a network is represented by the number of firms already linked to the network. Thus, in this chapter we mainly speak about *consumption network externalities*. The aim of our empirical exercise is to test:

a) what is the relative importance of network externalities for firms in their decision to join a network;
b) whether this aspect has more significance for specific industries and specific regions;
c) whether reasons exist at both the corporate and regional level explaining the different perspectives that firms have on consumption network externalities.

Our empirical investigation in this chapter is based on two methods. It contains first of all a *descriptive representation* of the results from our database, and is therefore useful to discover to what extent network externalities are the motive to join a network. On the basis of exploratory contingency table analysis we will be in a position to develop a behavioural analysis of the most important reasons for adoption. In particular, our analysis will concentrate on the identification of:

a) the main reasons for adoption;
b) the main reasons for non-adoption;
c) the main conditions for future adoption;
d) the main reasons for the dissatisfaction with the telecommunications technologies in use.

These reasons will be analysed in the case of the South and the North of Italy, in order to emphasise distinguishing features in *regional behaviour*. We expect *strong regional variations* in the results of our behavioural analysis. As mentioned in Chapter 6, it is anticipated that backward regions would not consider "the number of subscribers" as the main reason for the first adoption, as the adoption process in these areas is still in the first phases of diffusion. This means that a critical mass has not yet been achieved, especially because these new networks are mainly connecting regions in the South, and the link with other areas in Italy is not always possible. On the contrary, we expect that in advanced regions one of the main reasons for adoption is explained by "the number of already existing subscribers". This view stems mainly from two different elements: a) the different stage of diffusion these technologies have reached in the two different areas: that is, reasonably advanced adoption phases in central regions, in contrast with backward regions; b) the different economic environment in which firms are located. The same behavioural analysis is conducted at the *industry level*, separating the sample into firms belonging either to the industry or to the service sector.

The second step of the empirical analysis in this chapter is to obtain a further confirmation of the results through an *interpretative analysis*. In such an analysis the aim is to define which specific reasons explain the industry and regional variations by considering network externalities as the main general reason for adoption. The methodology used for our interpretative analysis is based on a discrete choice modelling approach, in particular on the basis of multinomial logit models, which will be conceptually described in Section 7.3. The choice of logit models has been driven by the essentially discrete nature of the database on the choice of a firm to join and use a network. On the basis of a series of explanatory variables which may reasonably explain the willingness to adopt (such as cost and tariff incentives, supply support, existence of a high number of firms already connected) logit models will be estimated, so that we will be able to indicate whether network externalities are considered as one of the main reasons for adoption.

The structure of this chapter is presented in Figure 7.1. The chapter provides both a descriptive and an interpretative analysis. Section 7.2 contains the descriptive results obtained by our empirical sample, underlining the regional and industry variations in the firms' replies. Section 7.3 explains the methodology for the interpretative analysis. Section 7.4 is devoted to the logit model estimation results, and finally Section 7.5 contains some concluding remarks.

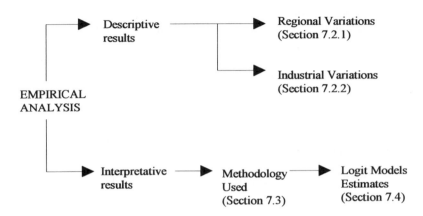

Figure 7.1. Structure of chapter 7

7.2. Network externalities as the main reasons for adoption: descriptive results

This section presents descriptive results, which are important for the definition of the empirical validity of the "economic and spatial symbiosis" framework. In particular, this section incorporates a *behavioural analysis* pinpointing the most important decisions to adopt telecommunications technologies. In this way, we are able to test whether our first research proposition interpreting network externalities as one of the major reasons for joining the network is empirically valid.

The present section provides contingency table analyses of questions related to the adoption decision process of new networks and services. The contingency tables show the percentage of firms replying positively to each specific choice, dividing the sample between the South and the North. Correlation analyses based on Chi-Square (X^2) test statistics and P-values[1] are run between each (positive) reply and the regional dimension. In this way we are able to present which replies are correlated with the regional dimension. The same analysis is run at the industry level. In the next sub-sections we present the descriptive results at the regional level (Section 7.2.1) and at the industry level (Section 7.2.2).

7.2.1. Regional variations

The results on the level of adoption show higher adoption rates in the North than in the South, especially in the case of infrastructure. Among infrastructures, packet switching networks and local area networks (LANs) prevail, while in the services, data banks and telefax are mostly adopted.

Concerning the results of our *sample in the South*, adoptions are overwhelmingly concentrated in 1991 and involve services much more frequently than infrastructures (Table 7.1). Among infrastructures, packet-switching networks and local area networks (LANs) are the only ones with a significant number of adoptions.

Table 7.1.
Adopted telecommunications network and services by macro-areas*

	SOUTH	NORTH	ENTIRE SAMPLE	CHISQ**	P-Value***
NETWORKS					
Optical Fibre Network	2.9	5.7	4.3	0.35	0.55
ISDN	2.9	5.7	4.3	0.35	0.55
Packet Switching Network	11.4	31.5	21.4	**0.15**	**0.04**
LAN	11.4	40.0	25.7	**7.48**	**0.00**
Broadband Network	8.6	5.7	7.1	0.21	0.64
SERVICES					
Videotex	22.8	34.3	28.6	1.1	0.3
Videoconference	2.8	2.8	2.8	0.0	1.0
Electronic Mail	51.4	28.6	40.0	**3.8**	**0.05**
Electronic Data Interchange	17.1	17.1	17.1	0.0	1.0
Databanks	71.4	54.3	22.8	2.2	0.1
Telefax	37.2	100	68.6	**32.1**	**0.0**

* These values represent the percentage of positive replies obtained for each question in the two areas

** The chi-square (CHISQ) represents the degree of dependency between the percentage of positive replies and the regional variables.
The results indicating a significant dependency are reported in bold characters

*** The P-value represents the probability to accept the null hypothesis when this is true
The P-value is acceptable up to a value of 0.05 (Loehlin, 1987; SAS, 1989)

This result may be explained by the fact that packet switching networks can only be used for data transmission because of their technical characteristics, and data transmission is the second reason for communication among firms, after voice. Among services, data banks prevail, followed by electronic mail in both service and manufacturing sectors. This means that both interactive services (such as electronic mail) and non-interactive services (such as data banks) have been well received. Ranking the importance between these two most popular services, data banks are ranked more highly than electronic mail. This is mirrored by actual use: despite highly individual patterns, on average data banks are used on a weekly basis, whereas electronic mail is only used occasionally.

The results from the sample in *the North of Italy* are quite different, as expected. Adoptions are much more distributed over the last 5 years. Among infrastructures, LANs prevail, followed by packet switching networks (Table 7.1). Their higher penetration rate can easily be attributed to two main reasons: a) these are typically data transmission networks, and transmission is a necessary aspect of these technologies; b) moreover, the technical elements associated with these networks were introduced in the market some years before the technical elements supporting other kinds of networks, i.e. ISDN or broadband networks. This fact has two consequences: i) the technology itself is more reliable; ii) the adoption process has had more time to take place.

Among services, data banks also prevail in the North (54,3%), followed by the videotex service (34.3%) (offering among others also a database system), and by electronic mail (28.6%). Video conference is not yet diffused, and, to a lesser extent, this is also true for the electronic data interchange (EDI) service (17.2%), which has a lower percentage of adopters than expected (Table 7.1).

Reasons for adoption. Interesting results stem from the main reasons for adoption especially if a comparison between the two macro-areas is carried out. The most important reasons for adoption in our sample are "the importance of the technologies for the business" and "low costs of implementation and use". However, once the analysis is carried out at a regional level, variations in the replies emerge.

In *the South of Italy* interesting results originate from the analysis, since they support our first research proposition, that the number of already existing subscribers is the main reason for adoption. Reasons put forward for adoption stress very strongly "the existence of promotion and demonstration centres", but recorded almost as highly are the reasons concerning "the importance of the technology for the corporate business" and "low usage and implementation costs" (Table 7.2). These replies bear witness to the difficulties of adoption processes, since:

Table 7.2.
Main reasons for adoption by macro-areas

	SOUTH	NORTH	ENTIRE SAMPLE	CHISQ*	P-VALUE*
High percentage of already networked subscribers in your region	5.7	20.0	12.8	3.18	0.074
High percentage of already networked subscribers in other regions	11.4	17.1	14.3	0.467	0.493
Importance of the network or service for your business	37.1	65.7	51.4	**0.719**	**0.017**
Your suppliers were networked or were using the same service	5.7	11.4	8.6	0.729	0.393
Other firms in the same sector were connected	8.6	14.3	11.4	0.565	0.452
No other advanced communication network or service available	11.4	5.7	8.6	0.729	0.393
Low costs of implementation	42.9	22.9	32.8	3.17	0.075
Low costs of use	45.7	17.1	31.4	**3.17**	**0,01**
Increasing awareness of these technologies through demonstration centres	40.0	5.7	22.8	**11.66**	**0.001**
Image effect	20.0	14.3	17.1	0.402	0.526
Others	31.4	14.3	22.8	2.92	0.088

* As in Table 7.1

a) help from the supply side in demonstrating and promoting innovation is of crucial importance. Greater possible connectivity deriving from greater availability is not in itself a sufficient element to give rise to adoption processes;
b) moreover, another main reason for adoption in the first phases of a diffusion process is "the importance of innovation for the business". In other words, if a linkage is shown between the technology and the business areas, there is a higher rate of success for an innovation;
c) also stimuli on the financial side become strategic, once the risks of failure of the adoption have to be borne by the adopters. Risks are of course higher during the first phases of adoption.

Another remarkable result is the *total lack* of any references to "the number of connected users, suppliers, buyers and competitors in the same or other regions". This result stresses once more the empirical plausibility of our testable hypotheses. In the first phases of the adoption process, such as in the case of the STAR programme, other reasons prevail in the decision to adopt rather than "the number of already existing subscribers". In this area, in fact, there is still too small a number of subscribers, which could not be expected to act as an attracting factor for potential subscribers.

The major reason put forward for adoption by firms located in *the North of Italy* is "the importance of these technologies for business" (more than 65.7% of replies), followed by "low costs of implementation" (22.9%) and use costs" (17.1%) and by "the high percentage of already existing subscribers in the region" and "in other regions" (37%) (Table 7.2 above). These replies confirm our expectations, since in the case of an area in an advanced stage of diffusion:

- the total number of already existing subscribers is a crucial motive for previous adoption, once these technologies are of any interest to the business and their costs are reasonable;
- in comparison with the South, where the percentage of responses declaring this motive is around 17%, the number of subscribers plays a more crucial role in the decision process to adopt in the North. In fact, given the more advanced stage of the diffusion process in the North, one can explain this result by reiterating the fact that the higher the number of subscribers (after the critical mass has been achieved), the more interesting the adoption and thus the "bandwagon effect" occurs;
- in advanced stages of diffusion, the help from the supply side, through the establishment of demonstration and promotion centres, does not play any role in the diffusion process (only 5.7% of replies);
- in any stage of diffusion price incentives (i.e. free of charge technologies in the South) play a significant role as mechanisms which help the adoption process.

Reasons for non-adoption. The importance of the number of adopters in the first phases of diffusion is also confirmed by another result obtained, i.e. the reasons put forward by firms for NON-adoption. The results from the all sample show that the most important reasons put forward by firms for NON-adoption are "the importance of the network or service for corporate business" and "low percentage of connected subscribers, suppliers and customers".

Regional variations are once more evident. In the South two such reasons prevail in particular: "the low percentage of connected local users" and "the irrelevance for corporate business". Other significant adoption bottlenecks include "the low percentage of non-local connected users and of connected suppliers and customers", "the use and adoption costs (including time costs before efficient connection is achieved)", and "the inability to understand and/or evaluate these technologies" (Table 7.3).

The "bandwagon effect" in the advanced stages of diffusion processes is confirmed by the results of the North. In the case of "the reasons for non-adoption" the difference between the North and the South is even greater than the case of "the reasons for adoption". While in the South the main reason for non-adoption was "the low percentage of regional subscribers" (71.4%), in the North the main reason for non-adoption has been identified as "the non-importance of these technologies for business" (48.6%), followed by "the high costs of use" (37.1%) (see Table 7.3). The reasons for non-adoption in the North have very little to do with very low levels of subscribers; they are linked rather to business interests and financial constraints.

Table 7.3
Main reasons for non-adoption by macro-areas

	SOUTH	NORTH	ENTIRE SAMPLE	CHISQ*	P-VALUE*
Low percentage of connected subscribers	71.4	14.3	42.8	**23.33**	0
Low percentage of connected subscribers in other regions	45.7	8.6	27.1	**12.21**	0
None of your suppliers or customers using them	57.1	31.4	44.3	4.69	0.03
You do not think they may be usefull for your business	51.4	48.6	50.0	0.057	0.81
You do not see their importance in your business	28.6	14.3	21.4	2.12	0.14
High costs of use	22.9	37.1	30.0	1.70	0.19
Other	28.6	14.3	21.4	2.12	0.14

* As in Table 7.1

Main conditions for future adoption. Table 7.4 presents the main conditions for future adoptions for both networks and services. Even from these results it is clear that "a higher number of subscribers" does not play an important role in stimulating adoption *in the North* (11.4% in general; 8.6% in the case of customers and suppliers), where the interest in future conditions for adoption is more linked to "the technical progress of networks" (20%), to "an increase in their efficiency" (14.3%) and "a reduction in the price of use for services" (11.4%).

Table 7.4
Main conditions for future adoption by macro-areas

NETWORKS	SOUTH	NORTH	ENTIRE SAMPLE	CHISQ*	P-Value*
A higher number of people connected	28.6	11.4	20.0	3.14	0.07
A higher number of suppliers and customers connected	25.7	8.6	17.1	**3.62**	**0.06**
A better geographical distribution of the network	11.4	5.7	8.6	0.73	0.39
A reduction in the price of use	11.4	11.4	11.4	0.0	1
A reduction in the price of access	8.6	8.6	8.6	0.0	1
More recent availability of the network	8.6	11.4	10.0	0.16	0.69
Technical progress in the network	25.7	20.0	22.8	0.32	0.57
Increase in the efficiency of the network	20.0	14.3	17.1	0.40	0.53
Good results from the previous adoption	14.2	5.7	10.0	1.43	0.23
SERVICES					
A higher number of people connected	40.0	14.3	27.1	**5.85**	**0.02**
A higher number of suppliers and customers connected	31.4	20.0	25.7	1.19	0.27
A better geographical distribution of the service	17.1	2.9	10.0	**3.97**	**0.05**
A reduction in the price of use	5.7	20.0	12.9	3.2	0.07
A reduction in the price of access	11.4	17.1	14.3	0.47	0.49
More recent availability of the service	11.4	5.7	8.6	0.73	0.39
Technical progress in the service	20.0	17.1	18.6	0.09	0.76
Increase in the efficiency of the service	25.7	17.1	21.4	0.76	0.38
Good results from the previous adoption	20.0	2.9	11.4	**5.08**	**0,02**

* As in Table 7.1

On the contrary, *in the South* the main condition for future adoption is related to "a higher number of subscribers", for networks (28.6%) and services (40%) (see Table 7.4). For services, the results are different than from the case of networks. While "the reduction in the price of use" still remains rather crucial in the North (20%), this has the same weight as "a higher number of suppliers and customers connected" (20%). "Technical progress" and "increase in the efficiency of the service" still play both a crucial role (17.1%). *In the South*, instead "a higher number of people connected" is stated to be the most important reason, especially in the case of advanced interactive services, such as videotex and electronic mail. For data banks, too, "a higher number of people connected" is the main expected condition for future adoption, explained by the fact that a higher number of people connected would assure a higher revenue to the service managers, and thus stimulate a better quality of the service (i.e. a larger variety of available information). For electronic mail and electronic data interchange services the most important reasons stimulating future adoption are also "technical progress" and "increase in network efficiency".

Reasons for dissatisfaction with the telecommunications technologies in use. The most interesting result is the regional variation in the reasons for the dissatisfaction with the quality of existing technologies. As expected, "the degree of satisfaction with the technologies" among subscribers in the North is higher than in the South (68.6% against 57.1%) (Table 7.5). Moreover, the main reason put forward for a high degree of dissatisfaction is "the too low number of subscribers" (46%) and "the too high costs of implementation" (30%).

Table 7.5
Degree of satisfaction of the quality of communication by macro-areas

	SOUTH	NORTH	ENTIRE SAMPLE	CHISQ*	P-Value*
YES	57.1	68.6	62.8	0.98	0.322
NO	42.9	31.4	37.1		

* As in Table 7.1

It is interesting to see that these two reasons are also very much dependent on the regional dimension. While in the North "the high costs of their implementation" are the most important reasons (54.5%) this does not apply to the South (13.3%) (see Table 7.6). Dissatisfaction does not stem from "the low number of subscribers" (only 9.1% of positive replies for this reason), as is the case in the South (73.3%).

Table 7.6
Major persistent user problems
in the case of dissatisfaction by macro-areas

	SOUTH	NORTH	ENTIRE SAMPLE	CHISQ*	P-VALUES*
Too high costs of implementation	13.3	54.5	30.7	5.06	0.024
They are no longer free charge services	20.0	0.0	11.5	2.48	0.11
The use of these technologies required profound organisational changes	0.0	9.1	3.8	1.42	0.23
Too low number of subscribers	73.3	9.1	46.1	10.53	0.001
Suppliers are not using these technologies	33.3	9.1	23.1	2.10	0.15
Competitors are not using these technologies	0.0	26.7	15.4	3.47	0.063
Other	54.5	20.0	34.6	3.35	0.067

* As in Table 7.1

The sample in the North shows a general greater interest in these technologies, witnessed by a high percentage of replies recording "the increased intensity of business relationships after the adoption of telecommunications technologies" (62.9% of replies in the North against 20% in the South) (see Table 7.7). For the 37% of "decreased business relationships after the introduction of these technologies", in the North the reasons justifying this reply are not linked to network externality effects. As Table 7.8 shows, none of the choices embodying a network externality effect (such as "no suppliers or customers using these technologies") are put forward as the main reasons for constant business relationships.

Again, these results witness the fact that at low penetration levels, such as in the South, "the low number of subscribers" becomes a constraint for the increase in the intensity of use of these technologies. In the case of more advanced adoption processes, the reason explaining the low level of intensity of use is "the non-importance for business", represented by 14.3% of replies.

Table 7.7
Intensity of relationships after telecommunications technology adoption by macro-areas

	SOUTH	NORTH	ENTIRE SAMPLE	CHISQ*	P-Value*
Increased intensity of business relationship	20.0	62.9	41.4	**13.25**	**0.0**
Constant intensity of business relationship	0	0	0		
Decreased intensity of business relationship	80.0	37.1	58.6		

*As in Table 7.1

Table 7.8
Main reasons in the case of unincreased intensity by macro-areas

	SOUTH	NORTH	ENTIRE SAMPLE	CHISQ*	P-Values*
None	20.0	11.4	15.7	0.97	0.32
Costs of initial investment	2.8	0.0	1.4	1.01	0.32
Lack of reliability on the technologies used	2.8	0.0	1.4	1.01	0.34
Level of telecommunication charges	0.0	0.0	0.0		
Lack of staff skills in their use	2.8	14.3	8.6	2.92	0.08
Services are not relevant to business	14.3	17.1	15.7	0.11	0.74
Your customers do not use them	0.0	31.4	15.7	**13.05**	**0.0**
Your suppliers do not use them	2.8	25.7	14.3	**7.45**	**0.006**
Too few subscribers in general	40.0	0.0	20.0	**17.5**	**0.0**
Restricted geographical diffusion of networks and services	8.6	2.8	5.71	1.06	0.30
Other reasons	8.6	5.7	7.14	0.21	0.64

* As Table 7.1

Concluding remarks. The results presented are quite satisfactory, since at least at a descriptive level, our expectations are fulfilled and our testable hypotheses verified with *strong regional variations*, as expected. The importance of the regional dimension in our replies is witnessed by the statistically significant level of the X^2 and of the P-value between the "network externality" variables and the regional dimension. It is surprising that a significant correlation exists for the regional dimension only for the network externality variables, as is represented by the contingency tables, where the statistically significant dependency is marked by bold print (Tables 7.2/7.8 above).

These results reinforce our expectations based on the general beliefs in the theoretical literature on consumption network externalities (see Chapter 2) that in the first phases of adoption of new networks and services (such as in the South), the main reason for non-adoption is "the low percentage of subscribers already linked to the network". It makes sense, therefore, to find that the main reasons for adoption in the very first phases of adoption process is *not* the number of already existing subscribers. Nevertheless, one of the major problems still existing in the use of new telecommunications technologies is the low number of subscribers. It should be noted, however, that all these results do not hold in the case of Northern Italy, where more advanced diffusion processes lead to opposite results, as expected.

7.2.2. Industrial variations

A completely different framework emerges when a descriptive analysis at the industry level is presented. A first consideration needs to be made here. The strong regional variations which were revealed by the regional analysis (presented in the previous section) are not at all so evident in the case of the industry analysis. This is witnessed by the very limited number of statistically dependent cases in the sectoral distribution of replies, as we will see below. Thus, the differences between industries do not seem to play a significant role in the explanation of the reasons for adoption.

The sample, divided between the industrial sector (both traditional and advanced sectors) and the service sector, shows only some small discrepancies in the replies concerning *the different reasons for adoption*, which can all be explained by the varying nature of the business structure of these economic activities. Among networks, packet switching networks and LANs remain the most adopted infrastructures for data transmission, while ISDN, broadband networks and optical fibre networks show a very limited number of subscribers, due to their recent introduction in the market. Concerning services, data banks and telefax mainly prevail, followed by electronic mail and then by the videotex. The very low videoconference adoption is entirely concentrated in the industry sector, while this technology is non-existent in the case of the service sector (Table 7.9).

Table 7.9
Adopted telecommunications networks and services by industry and service industry

	INDUSTRY	SERVICE INDUSTRY	ENTIRE SAMPLE	CHISQ*	P-Value*
NETWORKS					
Optical Fiber Network	2.7	6.1	4.3	0.48	0.49
ISDN	2.7	6.1	4.3	0.48	0.49
Packet Switching Network	18.9	24.2	21.4	0.29	0.59
LAN	24.3	27.3	25.7	0.08	0.79
Broadband Network	5.4	9.1	7.1	0.36	0.55
SERVICES					
Videotex	24.3	33.3	28.6	0.69	0.40
Videoconference	5.4	0.0	2.8	1.84	0.17
Electronic mail	48.6	30.3	40.0	2.45	0.12
Electronic Data Interchange	21.6	12.1	17.1	1.11	0.29
Databanks	64.6	60.6	62.8	0.14	0.71
Telefax	62.2	75.7	68.6	1.49	0.22

* As Table 7.1

Reasons for adoption. The main reasons for adoption show industry differences which can be easily attributed to sectoral characteristics (see Table 7.10). In the case of *the service sector*, "the importance of these technologies for business purposes" is to a large extent the most crucial reason for adoption (more than 66% of replies). This result demonstrates the importance of these technologies for the service sector. The number of subscribers does not represent an important reason.

In the case of *the industry sector*, "the low costs of implementation and use" are the main reason for adoption, followed by "the support provided by demonstration centres", i.e. by an active marketing policy. "The high number of subscribers in the network" as a reason to adopt has greater importance in the industry sector, than in the service sector.

This fact may easily be explained by two reasons:

a) more external relationships are established by firms belonging to the industry sector (with suppliers and customers). Service activities, however, are much more characterised by internal flows of information and thus the number of subscribers using the network is less important;

b) in most cases the service sector bases its internal flows of information on private networks whose adoption process, by definition, is not supported at all by network externality effects[2].

Table 7.10
Main reasons for adoption by industry and service industry

	INDUSTRY	SERVICE INDUSTRY	ENTIRE SAMPLE	CHISQ*	P-VALUE*
High percentage of already networked subscribers in your region	16.2	9.1	12.8	0.79	0.37
High percentage of already networked subscribers in other regions	18.9	9.1	14.3	1.37	0.24
Importance of the network or service for your business	37.8	66.6	51.4	5.80	0.02
Your suppliers were networked or were using the same service	10.8	6.1	8.6	0.50	0.48
Other firms in the same sector were connected	13.5	9.1	11.4	0.34	0.56
No other advanced communication network or service available	8.1	9.1	8.6	0.02	0.88
Low costs of implementation	40.5	24.2	32.9	2.10	0.15
Low costs of use	37.8	24.2	31.4	1.49	0.22
Increasing awareness of these technologies through demonstration centres	18.9	27.3	22.8	0.69	0.41
Image effect	16.2	18.2	17.1	0.05	0.83
Others	29.7	15.1	22.8	2.10	0.15

* As Table 7.1

Main conditions for future adoption. Concerning the main conditions for future adoption, "the number of subscribers" represents the most important reason for future adoption in the case of both networks and services. In the case of networks, also "technical progress" also plays an important role for future adoption. The industry sector is slightly more inclined to have a "higher number of people or suppliers and customers connected" than is the service sector. The structural features of business activities in the two macro-sectors can explain this result. In fact, the industry sector has considerably more contacts with other external agents, especially subscribers, while information in the service sector is more internal information. On the other hand for the service sector, the main conditions for future adoption are linked to the "technical characteristics of the network" and to "the increase in the efficiency of the network", as well as to "the number of firms connected".

Table 7.11
Main conditions for future adoption by industry and service industry

	INDUSTRY	SERVICE INDUSTRY	ENTIRE SAMPLE	CHISQ*	P-Value*
NETWORKS					
A higher number of people connected	12.8	15.1	20.0	0,917	0,388
A higher number of suppliers and customers connected	24.3	9.1	17.1	2,85	0,091
A better geographical distribution of the network	10.8	6.1	8.6	0,502	0,479
A reduction in the price of use	13.5	9.1	11.4	0,337	0,562
A reduction in the price of access	10.8	6.1	8.6	0,502	0,479
More recent availability of the network	13.5	6.1	10.0	1,077	0,299
Technical progress in the network	27.0	18.2	22.8	0,774	0,379
Increase in the efficiency of the network	18.9	15.1	17.1	0,174	0,676
Good results from the previous adoption	13.5	6.1	10.0	1,077	0,299
SERVICES					
A higher number of people connected	29.7	24.2	27.1	0.27	0.61
A higher number of suppliers and customers connected	27.0	24.2	25.7	0.07	0.79
A better geographical distribution of the service	13.5	6.1	10.0	1.08	0.30
A reduction in the price of use	10.8	15.1	12.8	0.29	0.59
A reduction in the price of access	13.5	15.1	14.3	0.04	0.84
More recent availability of the service	10.8	6.1	8.6	0.50	0.48
Technical progress in the service	18.9	18.2	18.6	0.00	0.94
Increase in the efficiency of the service	21.6	21.1	21.4	0.00	0.97
Good results from the previous adoption	16.2	6.1	11.4	1.78	0.18

* As Table 7.1

For advanced services, a higher number of subscribers represents the main reason for future adoption for both the industry and the service sector. This result again shows very limited *industry variations* in the replies (Table 7.11).

Intensity of relationships after telecommunications technologies adoption. The intensity of relationships after telecommunications technology adoption shows a strong sectoral dependency. The industry sector shows a very high decrease of business relationships (72.9% of replies), in contrast to only 42.4% for the service sector (Table 7.12). What is interesting is that the most important reason explaining this very high degree of decreased intensity of business relationship is represented by "network effects", i.e. by suppliers, customers and other firms in general not using them (Table 7.13). In the case of the service sector, the most important reason expressed is the irrelevant role of the service for the business.

Table 7.12
Intensity of relationships after telecommunications technology adoption by industry and service industry

	INDUSTRY	SERVICE INDUSTRY	ENTIRE SAMPLE	CHISQ*	P-Value*
Increased intensity of business relationships	27.0	57.6	41.4	6.71	0,01
Constant intensity of business relationships	0.0	0.0	0.0		
Decreased intensity of business relationships	73.0	42.4	58.6		

* As Table 7.1

The behavioural analysis at the industrial level has shown very limited industry variation in the responses obtained. What differences there are can be related to the business structure, characterised by:

a) the nature of information flows; the service sector is in general characterised much more by *internal flows of information*, while for the industry sector the strategic flows are *external flows of information*, with suppliers and customers;
b) internal flows are much more supported by *internal networks*, such as LANs (which in our sample are also the most adopted networks - 27.27% of replies - compared to all other possibilities). These networks, by

definition, are not based on network externality effects, as has been theoretically explained in Chapter 2 of the present study.

Table 7.13
Main reasons in the case of unincreased intensity by industry and service industry

	INDUSTRY	SERVICE INDUSTRY	ENTIRE SAMPLE	CHISQ*	P-Values*
None	16.2	15.1	15.7	0.01	0.90
Costs of initial investment	2.7	0.0	1.4	0.90	0.34
Lack of reliability on the technologies used	2.7	0.0	1.4	0.90	0.34
Level of telecommunication charges	0.0	0.0	0.0		
Lack of staff skills in their use	8.1	9.1	8.6	0.02	0.88
Services are not relevant to business	13.5	18.2	15.7	0.29	0.59
Your customers do not use them	29.7	0.0	15.7	11.6	0.00
Your suppliers do not use them	18.9	9.1	14.3	1.37	0.24
Too few subscribers in general	32.4	6.1	20.0	7.58	0.00
Restricted geographical diffusion of networks and services	8.1	3.0	5.7	0.83	0.36
Other reasons	8.6	5.6	7.1	1.59	0.21

* As Table 7.1

Our aim is now to see whether our results may also be explained through an interpretative analysis. First of all we explain the methodology used for our interpretative analysis (Section 7.3); Section 7.4 contains the results of the estimated logit model for the industrial and regional levels of analysis. Section 7.5 presents some concluding remarks.

7.3. Methodology for an interpretative analysis: the discrete choice modelling approach

The interpretative analysis of our first research proposition is based on a standard discrete choice modelling approach, with economic random utility theory as the underlying theoretical rationale and revealed preferences as the empirical orientation. Discrete choice models such as multinominal logit, nested multinominal logit, and multinominal probit models are now well-established model approaches that are applied in a wide range of fields[3].

The importance of these models for our analysis stems from the fact that in most cases the decision of a firm to join and to use a network is of a discrete nature; in our case, too, the behavioural analysis is based on revealed preferences and the database obtained may only be applied for discrete models.

The logit models in our empirical analysis will be based on the complete database. In the next section the willingness to join a network will be analysed, highlighting industry and regional variations in the decision to join the network. The results of the analysis go further than the exercise of testing our first research proposition. They also have policy implications, especially in the case of Southern Italy, where the results will be able to prove whether the financial effort made by the EC to promote the use of these technologies in less developed regions has, in fact, generated a willingness for future adoptions by local firms in the South.

A number of *explanatory variables* characterising the reasons to join a network have been selected, namely (Table 7.14):

Table 7.14
Definition of variables

	1	0
PRICE EFFECT	If low implementation and use costs have played an important role in previous adoptions	Otherwise
ROLE OF THE SUPPLY	If the existence of promotion and demonstration centres has played an important role in previous adoptions	Otherwise
BANDWAGON EFFECT	If the existence of a high number of users has played an important role in previous adoptions	Otherwise

a) the *price incentives* for a firm. Here a distinction is made between firms having replied that low implementation and use costs have been a crucial variable in the decision to adopt (PRICE=1), and firms which in previous

adoptions have not recognised low financial costs as a basic reason for adoptions (PRICE=0);
b) the *role of the supply* in supporting the adoption of new networks and services, through, for example, demonstration and promotion centres. The question here is whether firms have recognised the efforts made by the supply side as helpful in their decision-making process of previous adoptions (ROLE=1) or whether they have never taken the supply efforts into consideration (ROLE=0);
c) the *bandwagon effect* in the decision making process to adopt. In this respect, a distinction is made between firms which have recognised the number of adopters as a crucial variable in the decision making process for previous adoptions (NET=1) and firms which have assigned no role at all to the number of adopters in decision-making processes for previous adoptions (NET=0).

The choice of these explanatory variables has been based on a selection of a great number of potential explanatory variables, such as the size of firms, the sector firms belong to, the innovation capacity of firms, their flexibility with respect to changes, the importance of the technology for the business, learning processes, etc. Among all plausible categorical variables, we have chosen those having the highest degree of dependency with the dependent variable. The results of the dependency analysis are shown in Table 7.15. Among all possible categorical variables, it is interesting to underline that the *size of firms* and the *sector firms belong to* are both independent from the willingness to adopt[4]. This means that contact patterns do not differ between small- and medium-sized firms, and in addition the reasons for adoption do not vary among sectors. This second result is in line with what was found in the descriptive analysis at the sectoral level presented above.

The above-mentioned variables (points a-c above) represent the expected explanatory variable of the willingness to adopt. A measure of the willingness to adopt is given in our database by the revealed interest to adopt advanced (interactive) services in the near future. A 0-1 variable has thus been built on the distinction between firms which have revealed a preference for future adoption of advanced interactive services (i.e. electronic mail) (WILL=1) and those which have not (WILL=0).

A way of estimating logit models without incurring the risk of infinite solutions is to make particular assumptions on parameters. In our case, the models have been estimated by assuming that the sum total of the parameters over the various categories of all main and interaction effects is equal to zero[5].

The research strategy for estimating the above-mentioned logit model[6] was based on an initial estimation of the model including all main effects, except for the variables "regional dimension" and "sectoral dimension".

Table 7.15
Degree of dependency between the dependent variable and the categorical variables

Willingness to adopt*	CHISQ**	P-Values
PRICE	5.66	0.017
ROLE	3.81	0.066
NET	3.02	0.082

* Dependent variable

** CHISQ shows a certain degree of dependency between the categorical variables and the dependent variable

Besides the estimated parameter values for the main effects, also the P-values and the X^2 of the likelihood ratio will be presented in our results. Next the "regional dimension" and the "sectoral dimension" variables were introduced separately in the model. In this way, the extent to which these variables lead to a better fit of the model has been estimated, and the estimated parameter values are also presented in Section 7.4. below.

7.4. Estimated logit model for the regional and industry level of analysis

In this section we consider the estimated logit model with respect to the willingness to adopt in the near future at both the regional and industry level. In this way, the need to test the statistical importance of the sectoral and regional dimension has been fulfilled.

As stated in the previous section, in the tables presented in this section, it should be recalled that our logit models have been estimated under the assumption that the sum total of the parameters over the various categories of all main effects is equal to zero (Agresti, 1990). Taking the example of the price effect variable, it can be derived from Table 7.16 that the estimated parameter for the firms replying that in previous adoptions price incentives have been strategic for their adoption decision-making process - i.e. PRICE=1 - equals 0.49. Given the above-mentioned restrictions, the estimated parameter for firms which deny giving any role to financial support in their decision-making processes - i.e. PRICE=0 - then becomes -0.49.

Table 7.16
Estimated logit model with respect to the willingness to adopt at the regional level of analysis

Variable	Estimated Parameter	CHISQ	P-Value
Constant	0.135	0.09	0.762
PRICE	0.4913	3.29	0.069
ROLE	0.3879	1.28	0.257
NET	0.1037	0.07	0.793
REG	-0.3226	1.25	0.264

CHISQ = 4.18

P-Value = 0.6526

In discussing the results of the estimated logit model, we first concentrate on the main explanatory variables, leaving till the end any comments on the regional and sectoral dimensions.

From a *statistical point of view*, the results are rather satisfactory for the *estimated logit model at the regional level*. This model has a 0.65 probability value to explain the willingness to adopt. Moreover, the good fit of the model from a statistical point of view is also demonstrated by the low number of significant categorical variables and by the lack of significant interaction effects among these categorical variables.

The *economic interpretation* of the model is interesting. A first conclusion drawn from Table 7.16 is that firms having chosen "low implementation and use costs" as an important reason for adoption are also the most dynamic firms in terms of future adoption, as is shown by the positive sign of the estimated parameter of the PRICE variable. Consequently, *the financial incentives represent a very important stimulus* for future adoption. This result has very important policy implications, since it shows that in the first diffusion stages, the price variable plays an important role in the decision to adopt. Moreover, this result assumes even more importance when linked to the STAR programme. As mentioned in Chapter 6, the STAR programme provided these technologies free of charge and our analysis reveals how strategic this choice has been in stimulating a local demand for these technologies. However, this result also represents a useful lesson for future transitions to a commercial phase; i.e. a gradual move towards market prices is required in order to maintain and increase the adoption level achieved in the first instance with a financial incentive policy.

As far as the supply role is concerned, it is evident from our results that the existence of promotion and demonstration centres has positive effects on the willingness to adopt. This is indicated by the significant positive estimated parameter value with respect to the variable ROLE. This result stresses once again an established idea in the literature about the necessity for a *bridging mechanism* between demand and supply for the successful adoption processes of these technologies[7]. The profound adjustments in terms of technological and organisational changes required in order to adopt and exploit these new technologies are achieved only if technical and organisational support is provided by the supply side.

Another quite interesting result is presented by the existence of a "*bandwagon effect*" in the willingness to adopt. Contrary to what we expected, the number of already existing subscribers does not seem to stimulate future adoption, as is witnessed by the estimated parameter of the variable NET. The parameter of this variable, in fact, has a P-value of 0.79.

As far as the regional dimension is concerned, we introduced in our model a variable reflecting the location of firms (REG), assuming a value 1 when located in the North and 0 when located in the South. This variable assumes a negative value, thus underlining the fact that firms in the North of Italy have less willingness to adopt than the ones located in the South. This result is not surprising at all, since the North of Italy has higher adoption rates than the South, for those networks and services commercially available (see Table 7.1 above). Low adoption rates in the North are typical for those networks and services which are either in an experimental phase (such as ISDN or video-conference) or are still very limited in their geographical extension (such as fibre optic networks).

From the same estimated parameter (with opposite sign) we deduce a clear interest in future adoption for firms located in the South. This is an extremely positive result when we analyse it in the framework of the STAR programme. One of the aims put forward by the EC was to stimulate an interest for these technologies among firms in the South, and to show their importance for the business activities of these firms and for the future of these firms. This aim of the programme seems to have been achieved, as the sample in the South has demonstrated the positive attitude of Southern firms towards future adoption. However, before being sure of the positive results of the STAR programme, it is also necessary to test whether the other extremely important aim of the programme, i.e. an economic revitalisation of backward regions, has been achieved. This is the subject matter of Chapter 8.

Table 7.17
Estimated logit model with respect to the willingness to adopt at the industrial level of analysis

Variable	Estimated Parameter	CHISQ	P-Value
Constant	0.36	0.51	0.473
PRICE	0.74	6.4	0.011
ROLE	0.57	2.36	0.124
NET	0.03	0.01	0.939
SET	-0.86	4.91	0.027
ROLE * SET	-0.53	2.08	0.149

CHISQ = 7.23

P-Value = 0.405

At *a sectoral level of analysis*, the estimated logit model is less satisfactory, although rather interesting from an economic point of view. The statistically less satisfactory results in comparison with the regional case are witnessed by a P-value equal to 0.405 and by the presence of a statistically significant interaction effect (Table 7.17). In any case, the sectoral results confirm what was previously proved at a regional level. The financial incentives, as well as the supply support, explain quite clearly the willingness to adopt, while the "bandwagon effect" loses much of its explanatory effect.

As far as the *sectoral component* is concerned, it appears quite clearly that firms belonging to the service sectors are more in favour of future adoption, as is witnessed by a statistically significant negative estimated parameter for the industry sector. Moreover, the statistically significant interaction effects ROLE * SET demonstrates that especially those service firms supported by the supply are oriented towards future adoption.

The interpretative results obtained are satisfactory, although they do not seem to support our idea that *consumption network externalities play a role in the diffusion process of these technologies*. However, the results show that it is interesting to develop the analysis at a territorial level. In fact, the *regional dimension explains part of the innovative behaviour* of firms and it is not just another additional variable in an already complex interpretative framework.

The present behavioural analysis has some policy implications, especially in terms of successful innovation policies for developing a local innovative demand. This is only the first part of the innovative process and of an innovation policy. To be successful, an innovation policy is expected to generate positive results on the production side, stimulating productivity and economic growth. Concerning this aspect, we present empirical evidence in Chapter 8.

7.5. Conclusions

In this chapter we have focused on a behavioural analysis of the main reasons for adoption of advanced telecommunications technologies. In particular, we have presented both a descriptive analysis, through contingency tables analysis, and an interpretative analysis, through the estimation of logit models.

These two analyses have led to slightly different results. The descriptive analysis strengthens the role of the number of already existing subscribers as one of the main reasons for adoption, while the interpretative analysis shows "price mechanisms" and "help from supply side" as the major explanatory variables.

Both the descriptive and interpretative analyses have shown a *regional variation* in the results. Concerning the descriptive analysis, this shows a strong regional difference in consumption network externalities. In backward regions, where a critical mass has not yet been achieved, the reasons for adoption are not the number of already existing subscribers. On the contrary, in advanced regions, the number of adopters represents one of the most important reasons for adoption. As expected, the results are the opposite when dealing with the main conditions for future adoption. In backward regions, the most important reasons for future adoption lie in an increase in the number of subscribers connected, while for advanced regions other reasons, such as low implementation and use costs, are relevant for future adoption. Moreover, an interesting result obtained from the contingency tables analysis is the statistical dependency of consumption network externality variables with the regional dimension. This result is even more important when one remarks that consumption network externality variables are *the only ones* showing a statistical dependency with the regional dimension. This statistical dependency has not been found at a sectoral level.

The interpretative analysis does not confirm the importance of consumption network externalities as a crucial explanatory variable of the willingness to adopt. This analysis, run on the basis of an estimation of multinomial logit models, singles out two crucial variables explaining the willingness for future adoption, namely "price incentives" and "support from the supply side; "the number of already existing subscribers" does not seem to be an important explanatory variable of the willingness to adopt.

Even in the interpretative exercise, the regional dimension is important and underlines the fact that there is a positive linkage between firms located in the South and the willingness to adopt. The interpretative analysis also emphasises some important policy implications, since the strategic elements of future adoption are highlighted.

However, the results at the industrial level have been less satisfactory. The expected differences in the replies between the industrial and the service firms have not emerged in reality. These results have therefore acted as a disincentive to carry out a further analysis at the industry level in the next chapter, where the analysis is in fact developed only at the corporate and regional levels.

As far as the STAR programme is concerned, our analysis clearly demonstrates its successful results in stimulating the willingness to adopt in the near future. However, to evaluate its full degree of success, it is necessary to test its effects on the performance of (adopting) firms and regions. In other words, whether *production network externalities are exploited by firms and regions* must be tested empirically. This is our second hypothesis deduced from our "economic and spatial symbiosis" framework and its empirical analysis forms the subject matter of the next chapter.

Notes

1. The Chi-Square (CHISQ) represents the degree of dependency between two variables. In our case the two variables will be the percentage of positive replies obtained in our interviews and the regional variable. The P-value represents the probability to accept the null hypothesis when this is true (McFadden, 1983).
2. For the theoretical reasons behind this concept, see Chapter 2 of this book.
3. On theory and applications of logit models see, among others, Ben-Akiva and Lerman, 1985; Bishop et al., 1977; Camagni, 1992b; Domencich and McFadden, 1975; Fischer and Nijkamp, 1985; Fischer et al., 1992; Griguolo and Reggiani, 1985; Leonardi, 1985; Nijkamp et al., 1985.
4. The CHISQ and P-Values between the willingness to adopt and the sector variable assume respectively 2.02 and 0.15. CHISQ, and P-values between the willingness to adopt and the size of firms (measured as the level of

turnover) assume respectively values 1.11 and 0.77. Even in the case of the size of the firm measured as the number of employees, the results are not significant: CHISQ assumes a value of 3.69 and P-value of 0.29.
5. The assumption that the sum total of the parameters over the various categories of all main and interaction effects is equal to zero is one of the possible assumptions which can be made on the parameters of logit models, in order to be able to estimate them without incurring in the situation of infinite solutions (Agresti, 1990, pp. 92).
6. The logit model has been estimated with the use of the SAS-package. See SAS / STAT Manual, 1989. This package estimates logit models under the assumption that the sum total of the parameters over the various categories of all main and interaction effects is equal to zero.
7. On the concept of the bridging mechanisms between the demand and the supply side in the telecommunication sector, see a recent work by Camagni and Capello (1991).

8 Industrial and regional variations in production network externalities

8.1. Introduction

In this chapter we deal with our second research question, *whether firms and regions gain from the existence of network externalities*. As our "economic and spatial symbiosis" framework shows from a theoretical point of view, firms and regions are expected to gain from being linked to a network. As stated in our conceptual framework (see Chapter 4), the advantage a firm may achieve through the use of a telecommunications network or service is reflected in its performance. At the heart of this statement is the definition of network externalities, since the marginal benefits a firm gets from being networked are higher than the marginal costs it pays. Consequently, in this chapter the second basic research issue, identified in Chapter 2, will be analysed.

This chapter contains an extremely important issue, since it moves on to the measurement of *production network externalities*. All problems associated with the empirical analysis emerge in this chapter. In fact, the empirical test of what we argue theoretically is fraught with difficulties. One of the main problems is that, in order to determine the impact of network externalities on corporate and regional performance, it is necessary to have a reliable measuring rod of network externalities, on one side, and corporate and regional performance, on the other. Moreover, production functions are influenced by a large number of elements, which are similar to network externality effects, such as innovation effects and economies of scale effects, and disentangling specific network externality effects from all these other effects is not so simple.

The empirical analysis is carried out with the database presented in Chapter 6, i.e. small- and medium-sized firms, belonging to different sectors, located in both the North and the South of Italy. The first part of this chapter is devoted to the identification of the methodological approach to network externality measurement (Section 8.2), while the results of the empirical exercise are set out in subsequent sections. The empirical analysis is developed by taking into account both a measure of the adoption and a measure of the use of these technologies. We expect to have *strong differences* in the results we will obtain with the use of these two different measures. Indeed, we anticipate that the simple adoption of these technologies produces very modest effects on the production function. These technologies have to be intensively used in order to have positive effects on the production function. The analysis is therefore carried out at a *firm level* taking into account the *sectoral and regional dimensions*. Results will thus be presented for the entire database, and for the South and the North of Italy.

The structure of the chapter is outlined in Figure 8.1. Section 8.2 covers the methodology of analysis, where in particular the conceptual approach is highlighted. Section 8.3 presents the effects of the *adoption* of new telecommunications technologies on the production function, while Section 8.4 contains the results of the *use* of new technologies on the production function. Section 8.5 provides a descriptive analysis of the structural characteristics of firms which have different capacities to exploit network externalities. This last empirical exercise allows us to have a profile of the structural variables which are essential for the exploitation of network externalities. This is a first step towards the identification of the micro-conditions associated with this exploitation, which is the subject matter of Chapter 9.

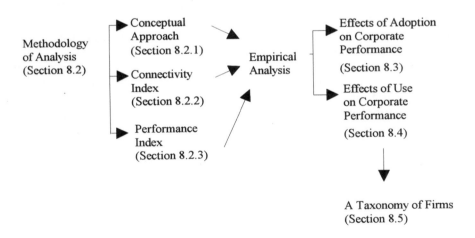

Figure 8.1. Structure of chapter 8

8.2. Methodology of analysis

8.2.1. The conceptual approach

Up to now in the present study the concept of network externalities has been explained in terms of the positive and increasingly intensive relation between the number of subscribers and the performance of firms. The higher the number of subscribers, the higher the interest for a firm to join a network, and thus the better the effects on its performance. In reality, this definition is far too broad to explain the concept of network externality. In a *static perspective* the interest of a firm is not to join the highest possible number of other firms connected via the network, but only the highest number of these firms directly or indirectly related to its own business activities. Thus, the decision to join the network is not simply related to the total number of firms already networked, but to the number of specific *business-linked firms* already present in the network. The most obvious reason for entering a network is, in reality, the possibility of contacting relevant groups such as suppliers, customers or horizontally related firms in a more efficient and quicker way.

Connectivity is in fact a measure of a linkage between two or more firms in a network. The economic connectivity measures the economic relationships among firms. When these relationships are pursued via a telecommunications network, then we can also speak of physical connectivity. What we argue here is that there is a strong relationship between these two kinds of connectivity; in particular *physical connectivity has no reason to exist if economic connectivity does not exist*.

Figure 8.2 is a schematic representation of physical connectivity with the use of graph theory. If, according to this theory, we represent firms as "nodes", or "vertices", and the physical linkages among them as "arches", or "edges", the outcome is a (undirected) graph of vertices and edges representing all potential physical communication (or contact) lines that firms can entertain among themselves.

As we have just mentioned, the real interest of a particular firm, in a static world, is not to be linked to all other possible subscribers, but to achieve full connectivity among only those firms related to its specific business. If we represent such firms in our graph with a bold vertex, and their economic relationships with other firms with bold edges, the real matrix of first order relationships will emerge. With this matrix it is possible to measure the proportion of real physical connectivity of a certain firm with regard to potential economic connectivity.

The physical connectivity is what generates network externalities. *If the benefits a firm receives from physical connectivity is an increasing function of connectivity itself, then positive network externalities exist*, a situation represented by the positive derivative of the benefit function (Figure 8.3). Thus, so far we have described a way of measuring network externalities under the assumption of a static world. Hence, Figure 8.3 represents a possible way of measuring *static network externalities*.

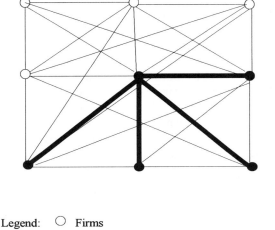

Legend: ○ Firms

● Business-linked firms

— Possible physical connections among firms

━ Physical and economic connections among specific "business-linked" firms

Figure 8.2. Undirected graph representing connectivity among firms

If we remove the assumption of a static situation, the potential linkages that firms are offered via a telecommunications network become of crucial importance. In fact, in *a dynamic perspective*, the interest of firms is not only to achieve static efficiency by developing better and quicker economic relationships with already existing suppliers and customers. The purpose and aim of networking are also related to the exploitation of other economic advantages, namely:

a) the achievement of new markets (Antonelli, 1992);
b) the development of product and process innovation, with the achievement of new and complementary technical, managerial and organisational know-

how (Capello and Williams, 1990; Fornengo, 1988; Rullani and Zanfei, 1988);

c) the control of the development trajectories of the strategic complementary know-how, by maintaining and increasing all strategic information sources (Camagni, 1991b);

d) the achievement of higher quality in the intermediate products provided by suppliers, and creating more competition among them by increasing their number (Capello and Williams, 1990).

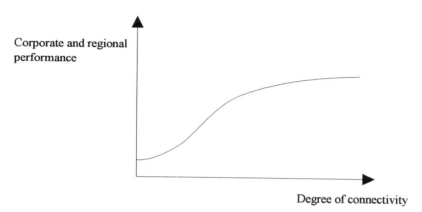

Figure 8.3. Increasing relationship between the degree of connectivity and the corporate and regional performance

In order to achieve these economic advantages, the increase in the degree of physical connectivity is the crucial vehicle for establishing new economic relationships, by achieving a higher degree of integration among economic agents. The exploitation of *dynamic network externalities* is in this case dependent on firms' expectations about the degree of "co-operation" or complementarity of other firms. If they expect that other firms will be willing to cooperate, then the degree of physical connectivity will increase, and, consequently, the benefits that firms receive from their connectivity.

With the use of this method, various important analytical questions remain from a methodological point of view. The *first open question* is related to the measurement of network externalities via a connectivity index which measures only direct connections. In other words, only *first order connectivity is measured via our method*, while second and higher order connectivity linkages are not taken into account. The choice of measuring a first order connectivity index requires in reality a careful choice. Second and third order connectivity loses the straightforward impact first order connectivity has on the production function, because the most strategic relationships which matter for the productivity of a firm are the direct relationships with suppliers and customers (Porter, 1990). Relationships among suppliers or customers of the same firm, representing what we call second and higher order connectivity for that firm, do not have the same direct relation with the performance of that firm.

A *second open question* which arises from the method we have presented to measure network externalities is that a connectivity index does not take into account the *intensity* of information flows. While in the case of the first question above we might disregard the importance of indirect connections of a firm on its performance, in the second case it is more difficult to avoid the problem. The intensity of use of a network, and not only its access, inevitably has an impact on corporate performance. Thus, any kind of connectivity index has to be adjusted in order to include a measure of the intensity of use. This problem will be taken into account in our empirical analysis. This point is also related to the problem of distinguishing the effects of simple adoption and of intense use of adoption.

The same approach can be applied to the regional level, by identifying a linkage between a connectivity index (measuring the relationships that firms located in that region have with other firms within and outside the region) and a regional performance index. Using the same logic as in the case of the firm level, the connectivity among firms located in that region can be measured with the use of graph theory. A positive derivative between these two indices would explain the existence of network externality effects.

A *third open question* of this method is that *the same weight in terms of economic importance is given to each link* although one can easily anticipate that each first order connection is certainly bound to be of strategic importance for the firm, which could otherwise easily refuse the contact.

The same three limitations as presented in the open questions above for the firm level, are also true for the regional level; again in this case, only first order relationships are taken into consideration, but the intensity of use of these technologies is missing in this approach. As already discussed above, the first and third open questions are not so crucial, since it can very well be that the most important connections influencing the performance of firms are direct connections and that all of them play a role in the performance of firms. The second open question is the most crucial, since the intensity of use is extremely important for our analysis.

The intensity of flows between regions may in principle be measured with the use of spatial interaction models, such as gravity models, designed with two principal nodes, and a factor measuring the obstacles raised by the distance between the two locations:

$$T_{ij} = K.W_i.W_j.f(c_{ij})$$

where:

T_{ij}: measure of the interaction between i and j
W_i: potential size node i
W_j: potential size node j
$f(c_{ij})$: cost of interaction between i and j

This model is normally defined at a meso/macro level. In the case of physical connectivity via a telecommunications network between two regions, the cost of interaction is represented by both entry costs and use costs of the network. These costs explain why, although by definition network externalities generate greater performance, there still might be a low intensity of use of a network. A logical explanation of this behaviour is represented by *transaction costs* which have to be borne, being measured in terms of entry costs and users' costs (psychological, financial, etc.). Shadow costs of the transaction, i.e. all costs which have to be borne in a transaction process towards the adoption and use of these technologies, can in this case be measured and used as a measure of costs of interaction between regions. This model might be helpful especially at the spatial network level, but it cannot be used for individual behaviour. In most cases, the decision of a firm to join and to use a network is of a discrete nature, so that - if there is a sufficient micro-database on firm behaviour - only discrete choice models can be used. Such models are compatible with spatial interaction models at a meso/macro level (see Nijkamp and Reggiani, 1992). Thus the gravity model approach offers a global framework of analysis, which can be empirically validated by using micro-based behavioural choice approaches.

8.2.2. The connectivity index

From what we said before, the *connectivity index* is of crucial importance for our analysis, since it is a measure of the degree of connectivity of firms. At a conceptual level, this index resembles the "N" variable in our mathematical exercise, which expresses the volume of information distributed among firms, depending on the number of contacts generated by different firms to be connected (see Chapter 5).

A connectivity index may be constructed with the help of graph theory (Andraisfai, 1977; Behzad and Chartrand, 1971; Christofides, 1975; Marshall, 1971). The major advantage of graph theory is that it does not presuppose a fully specified quantitative model; so it is a useful tool for qualitative behavioural models (Blommestein and Nijkamp, 1983). Some simple topological measures of network and graph structures may be useful for our purposes. They are concisely described now.

The *alpha-* (or redundancy) index gives information about the connectivity of networks. It is defined as follows:

$$\alpha = \frac{\mu}{(V-1)(V-2)}$$

where μ represents the cyclomatic number, defining the number of fundamental circuits in a network, i.e.:

$$\mu = E - V + G$$

where E is the number of edges (links) in the network, V the number of vertices (nodes), and G is the number of sub-graphs.

Since the α-index is defined as the ratio between the observed number of circuits and the maximum number of circuits, its value provides relatively sensitive information about the form of networks, both in the case of a single network and pairwise comparisons of networks.

The *beta*-index is defined as:

$$\beta = \frac{E}{V}$$

It can be considered as a simple measure of the complexity of networks. The β-index differentiates between simple topological structures (with a low β-value) and complicated structures (with a high β-value).

The *gamma*-index is calculated by dividing the actual number of edges by the maximum number of edges, i.e.

$$\gamma = \frac{E}{V(V-1)/2}$$

Like the α- index, γ provides information about the connectivity of the network. In fact, having the number of suppliers and customers of firms (representing the nodes) and the number of suppliers and customers connected to the firm with a computer network (representing the edges), the *gamma* index represents a simple measure of our connectivity index, measuring direct connections.

Unfortunately, our database does not allow the use of the gamma-index, since we lack quantitative information about the nodes of our networks, such as the number of connected suppliers and customers and the number of total suppliers and customers.

However, on the basis of the indications obtained by graph theory, a very simple connectivity index may be constructed, representing *the ratio between the number of real connections* to *the total number of potential connections*[1]. Although very simple in its formulation, it gives a measure of connection for each firm. The first open question mentioned before querying whether it was the right approach to build a connectivity index only on first order connectivity is not overcome by the way we build our index. However, we may be confident that second and third order connectivity has not the same effect that first order connectivity has on the production function.

The second open question regarding the lack of a measure of intensity of flows is rather important, since it also reflects at an empirical level the extremely important distinctions between the effects that a rare or an intense use of these technologies has on the production function of firms. We will also run the empirical analysis for a connectivity index weighted with the use of these technologies, thus taking into account their intensity of use. As we will see in the next sections, this index leads to different empirical results.

The third open question concerns the same weight given to all connections, or links. This limitation is not overcome by our connectivity index, although for our analysis it is not a strong limitation.

Despite the relative simple connectivity index used in the empirical analysis, the results obtained are rather satisfactory and provide evidence of what has been conceptually argued in our "economic and spatial symbiosis" framework. The effects on the regional performance are measured in our empirical analysis as the sum of the positive effects that all firms located in the regions receive. Thus, we postulate that the higher the number of firms enjoying network externalities in a regions, the higher the regional performance.

The research methodology followed in order to test the existence of production network externalities is based on a correlation analysis between the connectivity index and the performance index. The analysis contains an initial estimation of the correlation coefficient between the "row" connectivity index[2], which measures the simple adoption of these technologies by firms, and the performance index, at the national level. Subsequently, the extent to which the inclusion of the regional dimension leads to better correlation coefficients will be analysed. In this respect, *we may expect stronger regional variations in the results for the two different areas.*

The second step of the empirical analysis is devoted to the introduction of the "frequency of use" variable into our framework. Thus, instead of measuring the correlation between the degree of adoption of these technologies and the performance of firms, the analysis is run between the use of these technologies and the performance of firms. In the light of the theoretical reflections expressed in our conceptual framework (see Chapter 4), *we may expect stronger correlation in this case* than in the case of the simple adoption. It goes without saying that in this case also the regional dimension is introduced in the analysis, since significant regional variations are expected.

8.2.3. The performance index

The second index for the empirical analysis is the performance index. A very simple performance index was chosen, which represents the labour productivity of each firm, defined as the ratio between the turnover of firms in 1991 and the number of employees in the same year. This measure may vary according to specific features of firms, namely:

- the sectors firms belong to. In fact, there may be capital-intensive and labour-intensive sectors;
- the regions where firms are located. It might very well be that a sector is more productive in one region than in another because of the different regional penetration of innovation in capital and the different skill of the labour force.

To avoid any biased result with the use of our connectivity index, an analysis was undertaken on the database to see whether there was any consistent relationship between some firms' features and their productivity. In particular, an analysis was carried out to see whether the most "labour-intensive" firms belonged to a particular sector, or were located in a specific region; whether the largest firms were located in the same regions and in the same sector. The results of this analysis showed a completely random relationship among these variables. For this reason we have some confidence that the simple performance index measured as the "labour productivity" could be used in our analysis. The next sections are devoted to the empirical results regarding the existence of production network externalities.

8.3. Relationship between the adoption of telecommunications networks and the performance of firms and regions

In this section we present empirical evidence for our second research issue, i.e. *whether network externalities play a role in the performance of firms and regions*. In particular, in this section the main focus of the analysis is the identification of a possible correlation between the performance index and the connectivity index.

In light of our conceptual framework, we expect to find no correlation between the simple adoption of networks and services and the performance of firms. It is in fact not the simple connection to a network which generates benefits to a firm. It is rather the use of these technologies which creates production network externalities to the networked firms.

In order to test the first hypothesis deduced from our conceptual framework, the simple connectivity index described in the previous section was constructed, i.e. the ratio between the real number of connections to the number of potential connections for each firm of our sample. A very simple performance index has been chosen, representing the productivity of each firm and defined as the ratio between the turnover of firms in 1991 and the number of employees in the same year, as suggested in the previous section.

In order to be sure that the results are not biased by sector or size effects, the analysis has also been run taking into account the sector firms belong to and the size of firms. The "sector" variable has been introduced by running a multivariate correlation between the performance and connectivity indices and the sector firms belong to. The size variable has instead been taken into account by running the multivariate correlation in four different groups of firms with different size, in order to test whether the exploitation of network externalities was related to the dimension of firms[3]. If the size of firms has an impact, we expect an increasing (decreasing) value of the correlation coefficient when the size of firms increases (decreases). This methodology has been applied also at a regional level. Multivariate analysis with sector and size variables allow us to separate out network externality effects from more traditional effects of economies of scale and innovative processes. If there is any relationship between the level of connectivity and the performance of firms and this turns out to be independent from sector or size effects, variations in the performance of firms can be mainly attributed to the existence of (production) network externalities.

National Results. A correlation analysis was run on these two indices and interesting results emerged. Figure 8.4 presents a plot of observations of the performance index against the "row" connectivity index. It appears quite clearly from the plot that a very high dispersion exists in the way these indices are related in our sample and, thus, that a very low correlation exists between them. The Pearson correlation coefficient R confirms the first impression, having a value of only 0.069 (see Table 8.1). With this value we can go a step further by claiming that almost no correlation exists between these indices. *Our first hypothesis is thus confirmed, since the empirical analysis allows us to conclude that the simple adoption of these technologies as such has no effects on the performance of firms.*

This conclusion is similar to what other empirical studies have pointed out in the case of other advanced technologies.

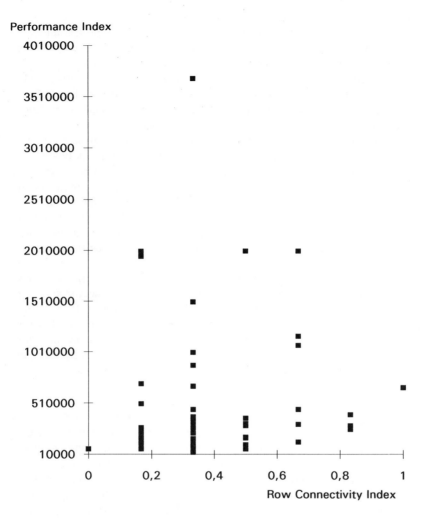

Figure 8.4. Relation between the row connectivity index and the performance index at national level

In particular, we refer to the studies on the diffusion of Computer Added Design (CAD) and Computer Added Manufacturing (CAM) technologies during the 1980s[4]. The results of these studies have pointed out that the simple adoption of these technologies as such did not lead to any increase in the performance of adopting firms.

Table 8.1
Correlation coefficients between the row connectivity index and the performance index by macro-areas

	NATIONAL RESULTS	
	Performance Index	Row Connectivity Index
Performance Index	1	0.069
Row Connectivity Index	0.069	1

	RESULTS FOR THE NORTH	
	Performance Index	Row Connectivity Index
Performance Index	1	0.398
Row Connectivity Index	0.398	1

	RESULTS FOR THE SOUTH	
	Performance Index	Row Connectivity Index
Performance Index	1	-0.0588
Row Connectivity Index	-0.0588	1

Also in the case of adoption processes of CAD/CAM technologies, the results at the empirical level have not fulfilled the expectations based on the high technological potentialities of these technologies. The impact of these technologies on the performance of firms has in fact been limited by economic and organisational elements, which have profoundly changed both the speed of adoption and the effects of the adoption on the performance of adopting firms.

Results do not change when the size and the sectoral variables are introduced in the analysis. The multivariate correlation analysis run introducing the sectoral variable leads to a similar result for the correlation coefficient, which assumes a value of 0.08. When the analysis is repeated in the four groups of firms with different size (in terms of both employment and size), the correlation coefficient changes randomly, and does not demonstrate any relation with the size of the firms.

Before focusing our attention on what happens when the frequency of use is introduced in the analysis, a rather provoking question is to see whether this result changes once the regional dimension is taken into consideration. *What we expect from the regional analysis is a greater capacity of advanced regions to exploit network externalities for a series of reasons*, which can be related to two different features:

a) different stages of diffusion of these technologies, which may explain: i) different levels of know-how guaranteeing the exploitation of network externalities; ii) different experience in innovation exploitation; iii) a different experience in organisational and managerial changes required to introduce these technologies;
b) a different economic environment which allows: i) a different specialised labour market; ii) a different "imitation effect" from successful adoptions of pioneering firms; iii) a different level of service sector for technical, organisational and managerial support; iv) a different degree of entrepreneurship able to deal with the risks which accompany all innovative processes; v) a different presence of flexible industrial structures able to accept the radical organisational changes required to exploit network externalities.

Regional Results. At the regional level our expectations appear to be verified. Figures 8.5 and 8.6 present the plot of the performance and connectivity indices, respectively in the case of the North and of the South of Italy. Figure 8.5 presents the results for the North of Italy. A higher degree of correlation appears there in comparison to the South (Figure 8.6), where the situation does not represent a clear trajectory. These impressions are sustained by the results of the Pearson correlation coefficients, changing to 0.398 for the North of Italy, and -0.058 for the South of Italy (see Table 8.1 before). The regional dimension is once again important in explaining the results of the empirical analysis. The national result is an average value of the two regional analyses, which separately show a different pattern. For the South the correlation is absent, with a value near zero and a negative sign. For the North of Italy, it is undoubtedly true that the situation improves achieving 0.39 as a correlation value and thus showing a weak correlation between the two indices. This result confirms our hypothesis of limited effect of adoption on the performance of firms. Results do not vary in the two regions, when the analysis is run taking into account the sector to which firms belong. In fact, the multivariate correlation analysis shows a similar correlation coefficient value: 0.4 in the North and 0.03 in the South. Moreover, the correlation run separately for the four groups of firms with different size does not show any clear relation between the dimension of firms and the correlation coefficient values.

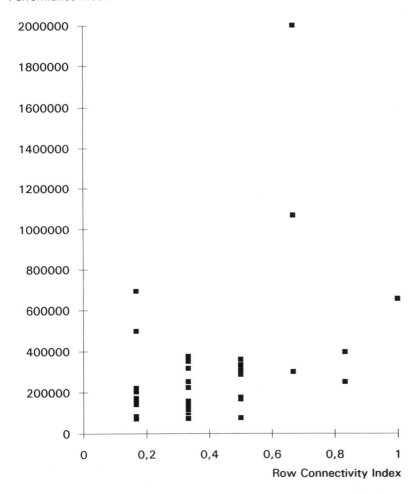

Figure 8.5. Relation between the row connectivity index and the performance index for the North of Italy

The next logical step is to see whether our second hypothesis is correct, i.e. whether an intense use of these technologies has an impact on the performance of firms. This is the subject matter of the next section.

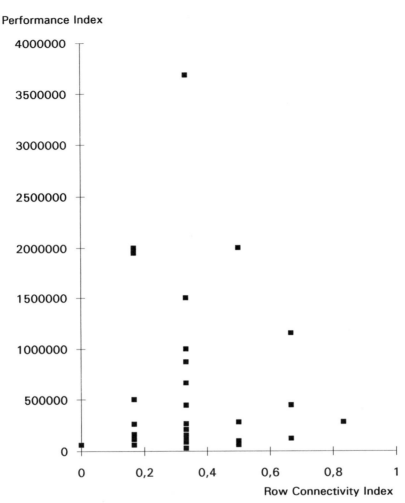

Figure 8.6. Relation between the row connectivity index and the performance index for the South of Italy

8.4. Relationship between the use of telecommunications networks and the performance of firms and regions

This section draws attention to the relationship between the use of advanced telecommunications technologies[5] and the performance of firms. The performance index remains the one constructed in the previous section. The connectivity index is instead adjusted to the frequency of use of these services.

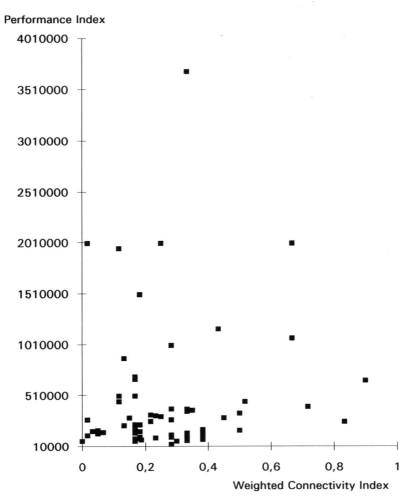

Figure 8.7. Relation between the weighted connectivity index and the performance index at national level

In this case we expect a strong correlation to exist between the performance and the connectivity index, since production network externalities can be exploited only if these technologies are used by adopters. In fact, more information and know-how are achieved via the network only when a systematic and strategic use of these technologies is put in place. As we have claimed several times in this work, the simple adoption is only a necessary but not a sufficient condition to achieve economic advantages. Moreover, we expect an even stronger regional variation in the way production network externalities are exploited, compared with the previous case of correlation between the simple adoption and the performance of firms. In fact, an intense and strategic use of telecommunications technologies highly depends on the characteristics distinguishing the two areas, i.e. the different stages of diffusion processes and the different economic development of the two areas. Especially entrepreneurship, which is a strategic resource assuring organisational, financial and managerial flexibility to cope with innovation processes, is of strategic importance in order to exploit production network externalities. This resource is by definition a scarce resource in backward regions, and thus adoption processes are expected to produce lower economic advantages.

National Results. The results of the correlation analysis at the *national level* of Italy are presented in Figure 8.7 above. The plot does not yet show a correlation between connectivity and the performance of firms. In fact, one group of four firms shows a high performance level, despite a very low connectivity level. Moreover, at least three firms are in an opposite situation, showing a high connectivity level despite a very low performance level.

A very high percentage of firms is clustered around low-medium levels of performance and connectivity. The Pearson correlation coefficient R still shows a very low value, viz. 0.11 (Table 8.2). Also in this case, the sectoral and the size variables turn out to be insignificant in the analysis.

Regional Results. In light of the previous remarks, the regional analysis is expected to add much to our interpretative analysis. The results are satisfactory in this respect. Figures 8.8 and 8.9 below represent the plot of the connectivity index (weighted for the frequency of use) and the performance index, respectively for the North and the South of Italy. It appears immediately that there is a better fit for a linear correlation in the case of the North than for the South. This impression is confirmed by large differences in the Pearson correlation coefficients, whose value varies from 0.085 for the South, to 0.473 for the North (see Table 8.2). These results show that:

- the regional variation in correlation analyses is even greater in the case of the correlation between the simple adoption and the firms performance. The national correlation value is nevertheless still very low, because it averages an even lower R in the case of the South and a higher value for the North;
- as expected, the most developed regions are also the ones which gain more from network externality effects, while backward regions are not yet able to achieve economic advantages from their adoption;
- the use of these technologies is strategic for the exploitation of production network externalities. In Northern Italy, where these technologies are used

more frequently, the economic advantages from their adoption is certainly higher than in Southern Italy.

Table 8.2
Correlation coefficients between the weighted connectivity index and the performance index by macro-areas

	NATIONAL RESULTS	
	Performance Index	Weighted Connectivity Index
Performance Index	1	0.11
eighted Connectivity Inde	0.11	1
	RESULTS FOR THE NORTH	
	Performance Index	Weighted Connectivity Index
Performance Index	1	0.473
eighted Connectivity Inde	0.473	1
	RESULTS FOR THE SOUTH	
	Performance Index	Weighted Connectivity Index
Performance Index	1	0.0856
eighted Connectivity Inde	0.0856	1

Even in this case, results are not affected by sector or size effects. The multivariate correlation analysis run between the connectivity and performance indices and the sectoral variables do not vary. In the case of the North of Italy, R assumes a value of 0.44 (instead of 0.47), while in the South of Italy it assumes a value of 0.01 (instead of 0.08). When the same correlation analysis is run in the four different classes according to the size of the firms, correlation coefficients change randomly.

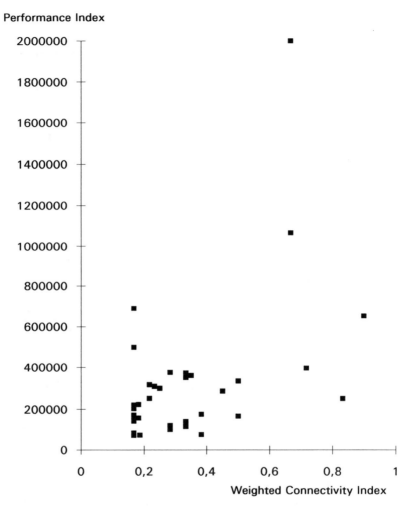

Figure 8.8. Relation between the weighted connectivity index and the performance index for the North of Italy

Figure 8.8 shows that in reality some outliers exist, i.e. there are a small number of cases with observed values that fail to confirm the model. In the presence of outliers, the well-known statistical method is the exclusion of the cases that do not belong to the average behaviour (Weisberg, 1980)[6].

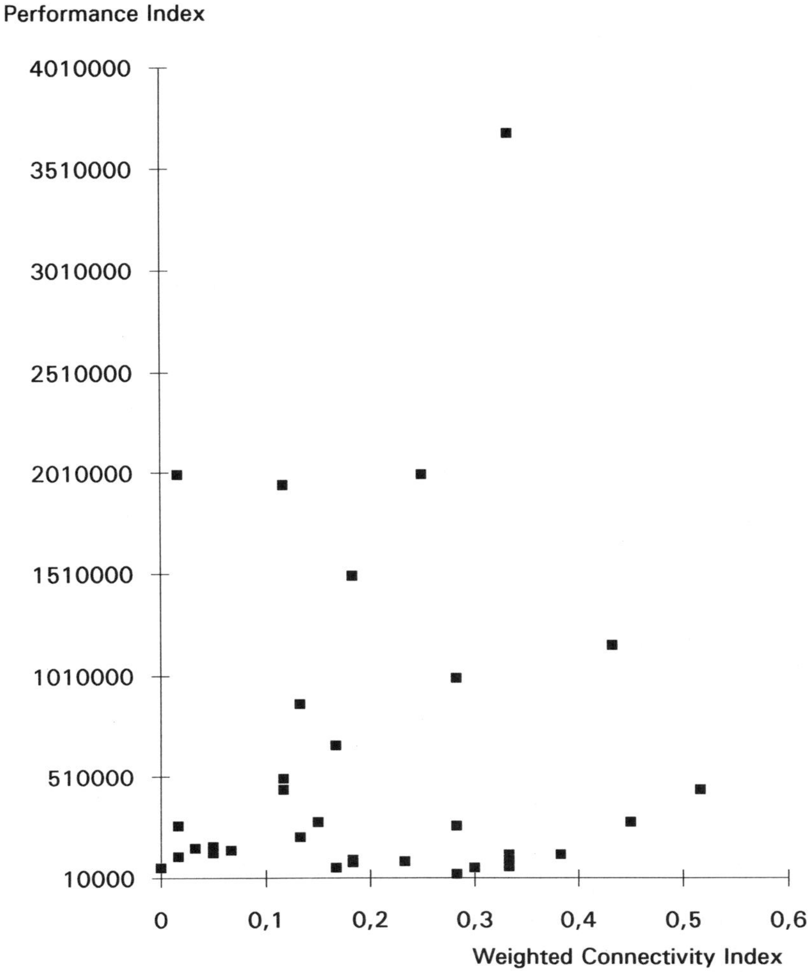

Performance Index

Figure 8.9. Relation between the weighted connectivity index and the performance index for the South of Italy

Following this strategy, the plot of the sample in the North shows a much clearer correlation pattern, witnessed by a Pearson correlation coefficient achieving a value of 0.57. A similar strategy applied to the sample in the South does not lead to a significant Pearson correlation coefficient, which even decreases to -0.068. Thus, it is impossible to argue that in the South production network externalities are exploited.

The results obtained in this part of the analysis are interesting. *Our hypothesis that firms and regions gain from network externalities has proved to be true in the case of advanced regions. Backward regions, on the contrary, seem to be quite unable to achieve economic advantages from the use of these technologies.*

This conclusion confirms what the vast body of literature on the spatial development of technologies claims (see Chapter 3). The general idea of this literature is that the diffusion processes of new technologies are governed by centripetal forces[7], i.e. tend to start in the centre, and develop towards the periphery. Our analysis reinforces these results and indicates that also the effects of the new technological development take place first in the centre and subsequently appear also in the periphery. This phenomenon may be explained by two reasons: a) the more advanced economic environment in which firms operate in more advanced regions, and which contains the capacities and know-how to exploit these technologies; b) the more advanced technological diffusion process in advanced regions which guarantees that the learning mechanisms take place.

The conclusion achieved in this section has some policy implications not only for the STAR programme, but also for all infrastructural interventions in backward regions. In fact, although this programme has achieved the aim of stimulating a local demand for advanced telecommunications networks and services (see Chapter 7), it has hardly generated any significant regional performance. Firms located in the South do not show an improvement in their business performance related to the adoption of new telecommunications technologies. The explanation of this negative result may be manifold:

a) an infrastructural programme, such as STAR, requires a long period in order to show its positive results in terms of the industrial and consequently regional performance. Thus, it may be too early to evaluate the regional effects that this programme has generated on the performance of firms, as the evaluation was undertaken only some months after the end of the programme;

b) as we have already mentioned, the impact of these technologies on the performance of firms and regions is far from being a linear and simple process. Many micro- and macro-conditions need to be present in order to generate an exploitation of production network externalities (see Chapter 6). The identification of the existence of these micro- and macro-conditions necessary to exploit production network externalities are the subject matter of the next chapter;

c) the negative judgement which could be given to the STAR programme following the results obtained in this part of the study is, nevertheless, compensated for by the positive results achieved in the empirical analysis on consumption network externalities, which was described earlier in Chapter

7. In fact, in this previous chapter, the interpretative analysis on the willingness to adopt these technologies showed a definitely positive attitude of Southern firms, even though they have not as yet achieved economic advantages from the adoption of these technologies. This situation, which could be seen as an illogical result, is in reality conveying a strong message. Firms located in the South have recognised the strategic role these technologies play for economic development, manifested through their willingness to adopt. However, they are still unable to exploit these opportunities given the low level of organisational and managerial know-how required. In other words, what is lacking is the entrepreneurship guaranteeing a successful innovation process. It is in this direction that future policy interventions should address their attention.

Before moving to our last empirical chapter, dealing with the micro- and macro-conditions supporting the exploitation of production network externalities, we turn to a description of the features characterising firms with different capacities of exploiting network externalities. This is the subject of the next section.

8.5. A taxonomy of firms: winners and losers

In the previous sections we highlighted the relationship between the degree of connectivity of each firm and the performance index. On the basis of these results we found out that certain groups of firms in our sample were much more able to exploit production network externalities, being characterised by higher levels of performance and connectivity. In this section our intention is to give a profile to these firms in terms of their structural characteristics which may explain the best way to exploit network externalities. For this purpose a cluster analysis has been run, with the aim of defining groups of firms characterised by structural similarities. The variables used in the cluster analysis are quantitative variables defining:

a) the *innovation capacity* of the firm, defined as the number of product, process, managerial and organisational innovations achieved in the last five years;
b) the *organisational flexibility* of the firm, defined as the number of functions which have faced organisational changes in the last five years;
c) the *business expansion capacity* of the firm. For this variable a dummy has been introduced, representing a 1 when the number of suppliers and customers had been expanded in the last five years, and a 0 otherwise;
d) the *sectoral specificity*, i.e. whether firms belong to the industry or to the service sector;

e) the *degree of connectivity* of the firm, used in the previous section, defined as the ratio of the amount of real connectivity to the potential connectivity, weighted for the intensity of use;
f) the *degree of productivity*, defined (as before) as the level of turnover by employee.

To run the cluster analysis on the basis of the previous variables the PROC CLUSTER[8] procedure in the SAS system has been used. This analysis optimises initially random centroids through a number of iterations to produce clusters of firms, all belonging to the same hierarchical level and with no overlapping members. Since the interpretation of clusters requires an ex-ante specification of the number of clusters, the analysis has been run following a two-step procedure. In the first step, the number of clusters has been defined on the basis of the output obtained by the cluster analysis run on standardised values of our variables with the centroid method[9]. In the second step, the clusters have been identified with a FASTCLUS[10] procedure. The analysis has been run separately for the North and the South of Italy, and interesting descriptive results have been obtained.

The aim of the exercise was to identify the structural similarities of these firms which had demonstrated similar behaviour in exploiting network externalities. For this reason, the results in the North, where a correlation between the productivity of firms and the degree of connectivity of firms exists, are much more satisfactory. On the contrary, in the South, where no correlation exists between the two indices, a cluster analysis around those two indicators does not lead to interesting results.

North of Italy. In the case of the North of Italy, six clusters on the six variables mentioned before have been identified (Table 8.3) which may be linked to the already defined correlation between the two indices of productivity and performance.

In particular, Figure 8.10 shows the cluster results plotted on the basis of performance and connectivity indices. It emerges quite clearly that:

a) a first cluster (cluster 3) represents the outlier case which also emerged in the previous section. This firm has a level of performance which is very far from the average levels of our sample, and can thus be treated as an outlier;
b) a second group of firms (cluster 5) represents those firms being able to exploit network externalities better and may for that reason be called *clear winners*. These firms are characterised by a high degree of innovation capability (much higher than the average values), a very high flexible organisation (above the average values) and a business expansion lower than the average. They also belong to the industry sector. Thus, firms which may be characterised as clear winners in the way they exploit

production network externalities are also characterised by high innovation capability and high organisational flexibility, but a low business expansion;

Table 8.3
Average population values for each observation in the North of Italy

	Cluster	Frequency	Innov. Capab.	Busin. Expan.	Conn. Index	Perfor. Index	Sector	Organ. Flex.	Nearest Cluster
LIKELY WINNERS	1	4	-0.47	-0.02	1.48	-0.1	0.5	0.09	6
LIKELY LOSERS	2	6	-0.39	0.45	-0.54	-0.15	0.66	1.57	4
OUTLIER	3	1	-0.49	-0.73	1.62	4.7	1	1.9	5
CLEAR LOSERS	4	13	-0.46	-0.62	-0.66	-0.27	0.3	-0.54	6
CLEAR WINNERS	5	2	2.13	-0.73	2.2	1.5	1	0.78	1
POTEN. WINNERS	6	9	0.72	0.85	-0.01	-0.3	0.22	-0.69	4

Legend:

Innov. Capab: Innovation Capability
Busin.Expan: Business Expansion
Conn. Index: Connectivity Index
Perform. Index: Performance Index
Organ. Flex.: Organisational Flexibility

c) a third group of firms (cluster 1), having high connectivity indices and relatively high performance indices (although lower than the average), may be labelled as *likely winners*. Firms having high connectivity indices, and a relatively high performance index are also characterised by an innovative capacity and a business expansion below the average, but a positive organisational flexibility. They belong to both the industry and the service sector (50%);
d) a similar cluster to the previous group of firms representing the likely winners is characterised by a cluster of *potential winners* (cluster 6). Firms belonging to this cluster have a performance and connectivity index below the average, but an innovation capacity and a business expansion above the

average. The organisational flexibility remains below the average for this group of firms, and they mostly belong to the service sector (78%). Statistically speaking, the nearest cluster to cluster 1 of likely winners is the cluster representing potential winners;

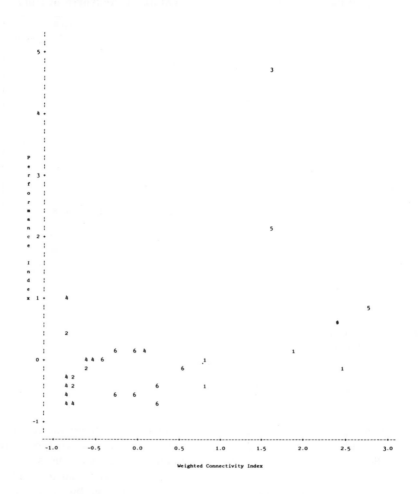

Figure 8.10. Description of clusters for the North of Italy

e) the fifth cluster (cluster 2) is characterised by firms with a very low connectivity index and a low performance index. These firms may be called *likely losers*. The levels of innovation capacity, organisational flexibility and business expansion are very much below the average. The cluster is formed mostly by firms belonging to the industry sector (66%). From a statistical point of view, the nearest cluster to the likely losers cluster is the one representing the clear losers;

f) the last group of firms is characterised by what we could call the *clear losers* since the levels of all variables are clearly below the average, especially the performance and connectivity index (Cluster 4). These firms mostly belong to the service sector (70%).

A general result is shown in this analysis. *Firms which may be labelled as winners in the way they exploit production network externalities (having high values of both the connectivity and the performance index) are also the ones having high levels of organisational flexibility and innovation capacity, characteristics stemming from high a degree of entrepreneurship (see Table 8.3). When these characteristics are lacking, the degree of exploitation of production network externalities is lower.*

These results confirm what has already been found in other studies on telecommunications diffusion[11]. In more qualitative studies the analysis on the diffusion processes of these technologies has pointed out the strategic importance of organisational flexibility and innovation capacities to achieve greater levels of performance via the use of these technologies. These results are not new in studies on diffusion processes of advanced and complex technologies and the manner of their exploitation. We refer here to the studies on the diffusion of factory automation technologies[12]: these works have also achieved the same conclusions, by claiming the importance of organisational flexibility and entrepreneurship in the diffusion and exploitation of these technologies.

South of Italy. As far as the South of Italy is concerned, the results are much less clear, as expected. In fact, the high degree of independence between the performance and the connectivity index does not allow us to explain the clusters on the basis of the dynamics of these two variables. Figure 8.11 presents the results for the South of Italy, where nine clusters have emerged. The difference between them does not lead to any logical explanation of the dynamics of the performance and the connectivity indices (Table 8.4). Some general remarks may be reported here, which show that no clear characteristics may be attributed to any dynamic behaviour followed by firms in terms of performance and connectivity indices:

a) two clusters are characterised by a performance and a connectivity index above the average: one is an outlier, having a performance and a connectivity index much higher than the average (cluster 9); the other group (cluster 5) is represented by lower degrees of innovation capacities and organisational flexibility than the average, while having a capacity for business expansion which is above the average;

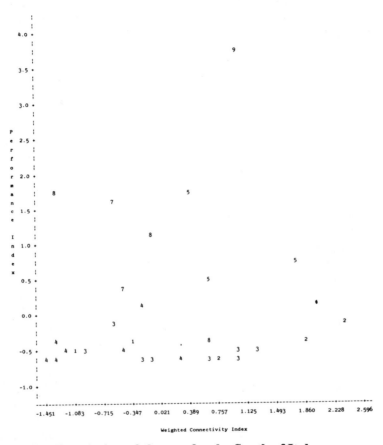

Figure 8.11. Description of clusters for the South of Italy

b) two clusters (clusters 2 and 3) are defined by connectivity indices above the average, but by performance indices below the average, and the behaviour of the other structural variables is completely the opposite in the two clusters;

Table 8.4
Average population values for each observation in the South of Italy

Cluster	Frequency	Innov. Capab.	Busin. Expan.	Conn. Index	Perfor. Index	Sector	Organ. Flex.	Nearest Cluster
1	4	-0.2	0.95	-1.08	-0.54	0	1.31	3
2	3	0.37	-1.01	2.35	-0.16	0	0.24	4
3	9	-0.15	0.95	1.24	-0.55	0	-0.82	5
4	9	-0.25	-1.01	-1.45	-0.63	0	-0.82	7
5	3	-0.25	0.95	0.38	1.74	1	-0.11	3
6	1	5.68	0.95	1	-0.59	0	-0.82	2
7	2	-0.25	-1.01	-0.59	1.68	0	-0.82	4
8	3	-0.1	-1.01	-0.1	1.12	1	2.75	1
9	1	-0.25	-1.01	1	3.8	1	-0.82	7

Busin.Expan: Business Expansion
Conn. Index: Connectivity Index
Perform. Index: Performance Index
Organ. Flex.: Organisational Flexibility

c) two clusters (clusters 7 and 8), characterised by higher performance indices but lower connectivity indices than the average, show similar features in terms of innovation capacity and business expansion (lower than the average), but opposite trends concerning the organisational flexibility;

d) two clusters (clusters 1 and 4) show both the performance and the connectivity indices lower than the average. For one of them all structural variables are below the average, while the other (cluster 1) shows an organisational flexibility and a business expansion higher than the average.

As mentioned before, these results are less interesting than in the case of the North, since no economic logic seems to emerge from the statistical exercise. This result is in part explained by the fact that no correlation exists between the performance and the connectivity indices, so therefore it is not surprising that no similarities may be found in terms of structural characteristics among firms which behave in a completely random way when dealing with production network externalities.

8.6. Conclusions

In this chapter our second group of hypotheses deduced from our conceptual "economic and spatial symbiosis" framework has been empirically investigated. In particular, we have concentrated on the effects of network externalities on the performance of firms and regions. At an empirical level, the identification of the existence of network externalities was fraught with difficulties, since the effects of network externalities had to be isolated from other elements positively affecting the production function of firms which could be easily confused with the network externality concept, such as the more traditional effects on the production function generated by innovation processes, or by economies of scale. The independence of our results from the sectoral and firms' dimension allows us to say that when there is a positive relationship between the connectivity and the performance index then it depends on network externality effects.

A conceptual approach to the measurement of network externalities has been formulated. In this respect, our attempt has been to keep the exercise as easy as possible, while at the same time trying to overcome the above-mentioned difficulties. The research strategy was based on the definition of two indices, one representing a measure of the performance of firms, the other being a measure of the level of physical connectivity achieved by each firm. In this way, if a positive relationship exists between these two indices, positive production network externalities are tested. These cannot be confused with advantages stemming from general innovation processes, or from "economies of scale", since the positive effects measured using this approach on the production function stem from the physical degree of connectivity of firms which is independent from the dimension of firms. As extensively explained in the theoretical part of this study, we do not expect a positive effect on the production function by the simple adoption of these technologies. These technologies generate their positive effects only if they are actually used by subscribers; in this way the network externality mechanisms can become apparent. Thus, we only expect a positive effect on the production function via an intense use of these technologies.

The results are satisfactory and hence our prior expectations are fulfilled. In fact, our empirical analysis has quite clearly demonstrated that a *simple adoption*, with very low use of these technologies, does not lead to much improvement in firms' performance, and consequently very low positive effects at the regional level are generated. Already at this stage of the analysis, regional variations emerge from the analysis, where a clear dichotomy in the behavioural patterns are manifested between backward and more advanced regions.

This regional discrepancy becomes quite evident once the analysis is run taking into consideration the *frequency of use* of these technologies. In this respect, our hypothesis is that a positive relationship exists between the use of these technologies and the performance of firms. If this correlation is confirmed, production network externalities are demonstrated at an empirical level. The results are positive only for advanced regions. The Pearson correlation coefficient shows in fact a positive correlation only in the case of advanced regions, while backward regions seem to manifest an incapability to exploit production network externalities.

These conclusions have policy implications for infrastructural interventions in backward regions. At first glance these conclusions may represent a sort of failure of the STAR programme itself, since the programme was run with the aim of encouraging economic development in the less favoured regions of the Community and in reality no regional economic improvement has been achieved. However, this result has to be examined in a more careful way. First of all, evaluations of regional impacts require *a long-term perspective*. Economic effects take place in the long-run, and only a long term evaluation can really test the benefits of the programme. The present empirical analysis was run only some months after the end of the programme, inevitably capturing little of the economic effects generated. Secondly, the impact of telecommunications technologies on regional development is not a straightforward mechanism. One of the greatest mistakes would be to expect a direct linkage between the supply of new technologies and economic and regional development. The link between these two elements, technology on one side and economic and regional development on the other, is a rather complex phenomenon. Its successful results stem mainly from a collection of essential elements which have to be present and have to be exploited in the right way.

The next chapter is devoted to testing empirically the existence of these crucial variables in order to identify whether the partial success of this programme may be related to the clear bottlenecks and barriers revealed in the causal path analysis of the "economic and spatial symbiosis" framework.

This chapter has also highlighted the structural similarities of our sample of firms. Especially for the North of Italy, the exercise has been quite useful since it allowed us to show the structural similarities of firms able to exploit production network externalities at different intensities. The results show that to be clear winners, a high degree of innovation capacity and of organisational flexibility is necessary. On the contrary, clear losers show a very low level of these two variables and also of business expansion. Different degrees in the levels of the structural variables characterise other clusters within the two extreme cases of winners and losers. The next chapter is designed in order to see whether these descriptive results are also true for an interpretative analysis of barriers and bottlenecks hampering the full exploitation of production network externalities.

Notes

1. The potential connection of a firm is defined as the total number of existing telecommunication services offered to firms.
2. In this study "row" connectivity index means the connectivity index constructed taking into account only the adoption data. This index is different from the so-called "weighted" connectivity index which is derived taking into consideration the "frequency of use" of adopted telecommunication services. This index was in fact constructed by multiplying the adoption data with a weight derived by the data on the frequency of use: a weight of 1 was given to services used every day, 0.7 to services used weekly, 0.3 for services used monthly and, finally, 0.1 for services used annually.
3. Four groups of firms were chosen according to Table 6.2: firms having a number of employees between 1 and 10 belonged to the first class; between 11 and 20 to the second class; between 21-50 to the third class; and over 51 to the fourth class. The exercise has also been run measuring size of firms on the basis of their turnover: 0-500 million lire the first; between 501 and 5,000 million lire the second, between 5,001 and 10,000 million lire the third and over 10,000 million lire the fourth.
4. For studies on the adoption of CAD/CAM technologies see, among others, Ayres and Miller, 1983; Camagni 1984 and 1985; Colombo and Mariotti, 1985; Coriat, 1981; Jelinek and Golhar, 1983.
5. Data on the intensity of use of advanced telecommunication services are available in our database.
6. For a detailed description of the outlier analysis, see Annex 3 at the end of the book.
7. See, among others, Boitani and Ciciotti, 1990; Bushwell and Lewis, 1970; Ewers and Allesch, 1990; Ewers and Wettman, 1980; Oakey et al. 1980.
8. For statistical detail on this procedure see the SAS Manual, 1989.
9. For a detailed description of this procedure see the methodological annex at the end of the book (Annex 4), where also the results of the cluster analysis are reported.
10. For the statistical details of this procedure see the SAS Manual, 1989.
11. See, among others, Camagni and Capello 1991; Fornengo, 1988; Rullani and Zanfei, 1988.
12. On the conditions for factory automation technologies see Camagni, 1985; Camagni and Pattarozzi, 1984; Arcangeli et al., 1987; Cainarca et al. 1989.

9 Micro- and Macro-conditions for production network externality exploitation

9.1. Introduction

In the previous chapter we tested the existence of production network externalities, by developing a correlation analysis between the level of connectivity and the level of performance of a firm. However, as already mentioned several times in this study, it would be misleading to postulate a linear relationship between the adoption of these technologies and corporate and regional development. The trajectory towards better economic performance as a result of the exploitation of network externalities is a complicated chain process, with many *micro- and macro-conditions* which have to be fulfilled in order to achieve a better economic and spatial performance. If these conditions are not fulfilled, *bottlenecks* in the exploitation of network externalities arise.

The aim of this chapter is to test a conceptual path analysis linking the use of new telecommunications technologies to better economic and regional performance, in order to identify the barriers and bottlenecks hampering the exploitation of production network externalities. In other words, the aim of the chapter is to "reveal the inner working of" the linkage between the connectivity and the performance index identified in Chapter 8, by defining the variables and the elements which characterise the relationship between the two indices of connectivity and performance.

The capacity to define these basic ingredients is of strategic importance in our work for two reasons. The first reason is that at present the analysis run in Chapter 8 on the correlation between the connectivity and the performance index does not tell us *why and under which conditions the correlation takes place*. Moreover, at present the results of Chapter 8 show the existence of a correlation between the two indices, without showing the *direction of causality*. We have interpreted the results as the higher the connectivity, the greater the performance, but this statement could easily be expressed in the reverse, by claiming that the greater the performance, the greater the connectivity. On the basis of the path analysis to be run in the present chapter, this ambiguity is overcome, by imposing and testing a clear causal path, starting from greater connectivity to better performance.

The second reason for the strategic role of this part of the study is the relatively great importance it has for policy implications, especially concerning the economic evaluation of the above-mentioned STAR programme in Southern Italy. In fact, once these bottlenecks and barriers are defined, it is easier to introduce an intervention policy with the clear aim of overcoming them and thus of achieving better industrial performance, and also, via multiplicative effects, better regional performance.

The importance of these conditions was clear at the very beginning of this work, when we stated (in Chapter 2) that the third research issue to be investigated was characterised by *the identification of those conditions under which network externalities were exploited*. Thus, the present chapter addresses the third group of hypotheses identified in Chapter 6, i.e. that production network externalities are exploited if:

a) a critical mass is achieved;
b) an innovative use of these technologies is achieved;
c) support from the supply side is achieved;
d) a certain degree of entrepreneurship exists, allowing the achievement of competitive advantage via these technologies;
e) a clear importance of these technologies for business purposes is envisaged by users.

As already mentioned in Chapter 6, at the macro level we expect production network externalities to be exploited by more advanced regions, since the five conditions expressed above are undoubtedly much more in evidence in developed regions.

This chapter contains the conceptual framework on which our causal path analysis has been built (Section 9.2). Section 9.3 describes the methodology used to test our Causal Path Analysis, while Section 9.4 the lists variables used for testing the model. The results are again presented for both the South and the North of Italy in Section 9.5 and 9.6, respectively. Finally, Section 9.7 contains some concluding remarks.

9.2. The conceptual model

The logic on which our conceptual model is built is based on the effects that a *supply-driven* programme for telecommunications development is expected to generate on the performance of firms and regions. The link between these two elements, technology on one side and regional development on the other, is a rather complex phenomenon. The successful results of this programme stem from a series of crucial elements which have to be present and have to be exploited in the right way. Figure 9.1 represents a logical structure of a regional impact analysis of telecommunications technologies development.

A typical *supply-driven development* programme is based on an injection of funds for the economic restructuring of specific areas. The first consequence on the supply side is the achievement of *greater possible connectivity* of these areas to more advanced telecommunications networks and services. The presence of more advanced telecommunications networks and services plays the role of new locational advantage in the area with consequent attraction of new firms to these areas. The arrival of new firms in the areas leads in the long term to better regional performance.

On *the demand side*, however, there are other mechanisms that, once they are set in motion, result in the achievement of competitive advantage and thus to greater competitiveness of firms and of the local economy. Moreover, if firms intensively exploit these technologies, they bring into the area information and know-how which is not present locally. An increase of such scarce resources helps once again to increase local competitiveness and to achieve better regional development.

The increase in the business performance can be analysed in terms of increased *efficiency*, greater *effectiveness* and enhanced *competitive advantage*. Increased efficiency is achieved, for example, by reducing costs and maintaining existing output levels through the use of the technology as a substitute for other inputs (i.e. clerical staff), thus maintaining the same organisational structure. Effectiveness is concerned with the capacity to deliver more and improved products within the existing resource base achieved through some marginal changes in the organisational structure. Competitive advantage is obtained through the exploitation of telecommunications technologies to achieve more strategic information and to generate product, process and managerial innovations under the conditions of radical organisational changes (Williams, 1987). The development of the innovative and strategic use of telecommunications technologies, generating positive effects on business performance, is strongly associated with profound organisational changes. In fact, innovative use of these technologies implies the inter-relation of technology and organisation as two unseparable variables (Mansell, 1990; Zeleny, 1985). Technologies in themselves appear as neutral devices, as a pool of opportunities available at a given cost, and can be interpreted as quasi-public goods. But what really matters, and what is not at all a public good, is the cultural and organisational capability of exploiting their potentialities, through a creative blending of technological devices, organisational styles and business ideas.

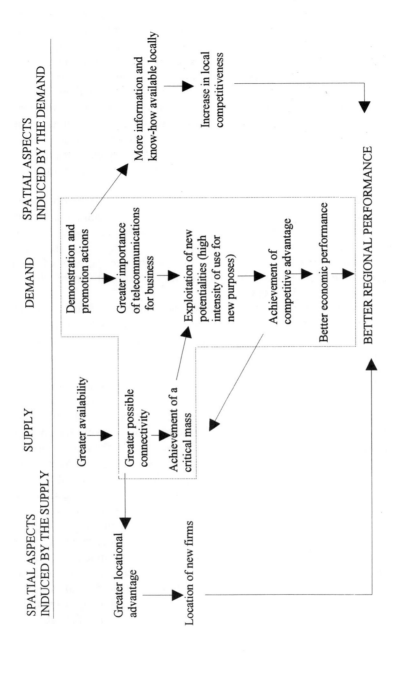

Figure 9.1. Causal path analysis of an economic and spatial symbiosis framework

This aspect becomes clearer by conceptualising the firm in terms of its transactional structure (Coase, 1937; Williamson, 1975). From this perspective, the adoption of new telecommunications technologies has to be seen within the context of an existing array of transactions, both within and between firms, which shape the nature of the organisation, particularly the division of functions and of labour (Capello and Williams, 1992). In particular, these organisational changes manifest themselves in processes of functional integration, the creation of new functions, changes in the relative importance of simple functions in the decision-making process, such changes being profoundly dependent on the existence of new telecommunications technologies governing transactions.

Table 9.1
Organisational changes at the intra-corporate, inter-corporate and spatial level

Organisational Changes	Intra-Corporate Level	Inter-Corporate Level	Spatial Level
Changes in the Division of Labour	* Requalification of Staff * Integration of Previously Differentiated Jobs * Changes in the Relative Importance of Jobs *Automation and Routinisation of Activities *Creation/Elimination of Activities	*Automation and Routinisation of Jobs *Elimination of Jobs	*Decentralisation/ Centralisation of Responsibilities
Changes in the Division of Functions	*Functional Integration *Changes in the Relative Importance of Simple Functions in the Decision-Making Process *Creation of New Functions	*Functional Integration *Internalisation/ Externalisation of Functions	*Decentralisation/ Centralisation of Functions *New Spatial Location of Functions

Source: Capello and Williams, 1992

At the same time, telecommunications technologies can generate requalification of staff, integration of previously differentiated jobs, changes in the relative importance of jobs, automation and routinisation of activities. A summary of the organisational changes is included in Table 9.1, which presents distinct organisational dynamics at an intra-corporate, inter-corporate, and spatial level. Thus, an innovative use is only achieved when these technologies are tailored to each adopter's needs (Camagni and Capello, 1991).

Learning processes and previous experience thus become strategic factors that, once they are present, facilitate the diffusion process. One may also argue that through the development and use of these technologies, users acquire a steadily growing awareness of the strategic relevance of these infrastructures for their day-to-day activities. Over time they develop the capacities to use these systems and, moreover, the capacity to specify their needs.

The link between greater availability and better regional performance is thus based on an extremely complicated mechanism of interrelated factors. If one of these factors is not present, the regional impact of new technologies does not take place, or at least is likely to be lower than expected. These conditions will only emerge after a profound analysis of the nature of these technologies and of the complex mechanisms accompanying the diffusion processes of these technologies.

On the basis of our conceptual approach, some *conditions* may be foreseen, which develop to avoid bottlenecks and barriers to the exploitation of production network externalities, on both the supply and demand side. Figure 9.2 summarises our conceptual model for production network externalities. The upper part of the chart represents the conditions on the demand side in order to exploit better production network externalities, while the lower part summarises the conditions on the supply side mentioned in Figure 9.1. When these conditions are present, we expect firms to be able to exploit the advantages from these technologies. This chart is a schematic representation of the dotted part of Figure 9.1, where the "spatial aspects" have been excluded. This exclusion has nothing to do with the logic of the chart, since it is perfectly true that locational factors and increase of scarce resources generate directly positive effects on the regional performance. However, for the sake of compatibility with previous empirical analyses, the effects at the regional level are analysed purely as the result of multiplicative mechanisms generated by firms located in the same area. Thus, in our model the analysis is also run at the micro-level in the two different areas under consideration, with the expectation of strong regional variations in the empirical results.

Figure 9.2 contains the structure of the causal path analysis we will test, which represents the dotted part of Figure 9.1. It comprises three blocks of variables. On the right hand side is the *endogenous variable* (block C), i.e. that variable which in the end has to be explained by all other variables in the model. The remaining variables in the model can be subdivided between *exogenous variables* and *intermediate variables*, respectively block A and B. The difference between these blocks of variables is the following. Variation in the exogenous variables is taken for granted and is not accounted for by the other variables in the model. Intermediate variables can instead be influenced by variation in the exogenous variables. Our model is quite simple, since it contains only one exogenous variable (the connectivity index); thus, no interrelations between exogenous variables exist. Moreover, our model does not contain feedback effects, and this again simplifies the analysis. Finally, we only have indirect effects between the exogenous and endogenous variables.

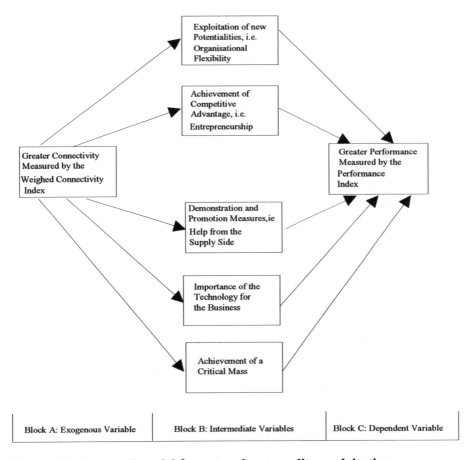

Figure 9.2. A general model for network externality exploitation estimates

The simple structure of our model reflects the aim of the exercise. Our intention in this chapter is to explain the relationship between the connectivity index and the performance index. Other variables may have been taken into account, such as the size of the firms or the sector they belong to, which may have a direct or an indirect effect on either the performance or the connectivity index. However, we restricted our analysis to some specific variables (presented below) for a number of reasons:

a) from the theoretical point of view, the variables used in our model represent the most common variables mentioned in the literature on telecommunications diffusion processes as those variables having strong influence on the decision to adopt these new technologies;
b) again from a theoretical point of view, there is no reason why larger firms should have different contact patterns than small firms, as is also witnessed by the results obtained in Chapter 7 and 8, where the size variable was insignificant with respect to the estimation of consumption and production network externalities. Larger firms may be more able to accept innovation, as is well known in theory, and this aspect is indirectly taken into account in one of the included variables in our model, namely the variable reflecting the innovative behaviour of a firm;
c) from a statistical point of view, a model with too many explanatory variables has low explanatory power. Thus, only the variables quoted in the literature as the most important variables explaining adoption processes were introduced in the model. Moreover, the good results achieved in the empirical analysis suggest that no important explanatory variable is missing (see Section 9.5).

As we mentioned several times in this study, we exclude any direct effect existing between the simple adoption of new telecommunications technologies and the performance of firms. The intermediate variables assume a rather strategic role, since they explain the correlation between the endogenous and the exogenous variables.

A very important assumption in path analysis is the recursiveness of the relation: relations between variables must have an evident causal direction and this allows us to disentangle the linkage between the connectivity and the performance index. As we mentioned already, the intermediate variables explain the conditions under which the correlation between the connectivity and the performance index takes place. In particular, their presence guarantees, as we conceptually explained earlier, that the adoption of new technologies generates better corporate performance. To be sure that the direction of causality between the performance and the connectivity index was the most appropriate one, we estimated the causal path analysis model also by imposing the reverse direction of causality that the one presented in Figure 9.2, i.e. a model where greater performance was the independent variable which explains greater connectivity. The poor results obtained have demonstrated once more that our hypothesis was the right one.

Concerning the supply side, at least three conditions have to be present in order to allow firms to exploit production network externalities:

a) the *achievement of a critical mass* of adopters, especially for interrelated services, e.g. electronic mail. The user value of these technologies is in fact related to the number of already existing subscribers. Our idea, although not entirely confirmed in Chapter 7, is that the number of already existing subscribers is one of the most important reasons for joining networks and services. The existence of at least a certain level of subscribers is a necessary condition to stimulate a *cumulative self-sustained mechanism*. If a critical mass is not achieved, the risk is that potential adopters are not sufficiently stimulated to adopt these technologies. In other words, the adoption process of these technologies has to be strongly supported by the supply until a critical mass is achieved;

b) another important factor is *constant assistance* in the first phases of development from the supply side, in terms of the technical, managerial and organisational support necessary for an innovative exploitation of these technologies. This process requires a strong organisational effort by the subscriber, who most of the time needs organisational support from people with expertise and experience. The STAR programme itself has certainly taken this aspect into account. Some specific measures were devoted to the implementation of "demonstration and promotion actions", through specialised centres whose task was to develop a "telematics culture" among potential users (see Chapter 6). However, in some countries such as Italy, these centres have put the emphasis of their work more on *technical* rather than *organisational aspects*. Centres in fact promoted these services, by mainly concentrating on their technical capacities, but while they assured technical support to users, they did not provide adequate information and advice about organisational problems and changes which have to be coped with in order to exploit production network externalities. We expect therefore that the lack of organisational support will act as a bottleneck in the exploitation of network externalities;

c) another crucial factor for the exploitation of production network externalities is the *clear identification of the way these technologies may be useful for business purposes*. As we explained before from a conceptual point of view, these technologies are multi-faceted complex technologies and require a certain degree of organisational changes in order to become useful to business needs. For these reasons, the supply side has to demonstrate the importance of these technologies for business needs, in order to stimulate an interest on the demand side.

Although there is a set of critical success conditions required on the supply side, it has to be added that also on the demand side there are some necessary requirements in order to exploit production network externalities:

d) to begin with, as a consequence of what has been said about the complexity of these technologies, the capacity of firms to accept new technologies and to exploit them is a very important consideration. We expect that firms have to be highly *flexible with regard to organisational changes* if they want to exploit production network externalities. The better the capacity of adapting the organisational structure to external changes, the greater the probability of exploiting production network externalities;
e) another factor which greatly assists firms to exploit production network externalities is their *innovative behaviour* or their level of entrepreneurship, this business acumen guarantees the achievement of competitive advantage through these technologies. In fact, via learning processes and previous experience of innovation, the firm may have acquired the necessary organisational and managerial know-how to be able to implement organisational changes required to achieve competitive advantage. The higher the number of innovations already adopted by the firm, the greater the probability of exploiting production network externalities.

In the next sections our aim is to test whether our causal path analysis is empirically valid. More specifically, Section 9.3 presents the methodology used, Section 9.4 provides the results obtained and Section 9.5 contains some concluding remarks.

9.3. Methodology of analysis: the causal path analysis

In order to test the *conditions* under which production network externalities are exploited, we used a methodology based on "causal path analysis"[1]. As we mentioned before, in this type of analysis the model is formulated as a path diagram, in which arrows connecting variables form the structure for defining reaction parameters which are essentially regression coefficients. Causal path analysis allows the use of hypothetical latent variables in the model. In particular, it is usually based on a conceptual arrow scheme depicting the relationships between latent variables and manifest variables[2]. This method is a combination of separate multiple regression equations within one framework.

Among various more or less similar statistical methods[3], the Generalised Least-Square (GLS) method has been chosen[4]. This method, like all similar methods such as a maximum likelihood estimation or partial least-squares method, allows us to construct a model of linear equations, where a given variable may be a dependent variable in one equation and an independent variable in another, so that it is able to maintain the difference between latent and observed variables. Thus, a GLS estimation procedure aims at estimating the various parameter values between latent variables and manifest variables.

The GLS is obviously a *least squares oriented* approach. This method allows the estimation of the regression coefficients in a simultaneous regression model. If each variable has been standardised to unit variance and mean zero, the value assumed by individual parameters represents the *order of magnitude of each particular independent variable* in explaining the dependent variable. The statistical significance of each parameter is given by the values of T-Student tests run in parallel to the coefficient estimation analysis.

Causal path analysis allows the interpretation of complex interrelated models, in which even feedback effects may take place and may be measured. Our conceptual model is, in fact, rather a simple one, since no feedback effects are taken into consideration and even the number of variables is limited. Nevertheless, despite the simplicity of the model, we preferred the use of causal path analysis rather than simple regression analysis. In fact, the *multidimensional characteristic* of our model would have been underestimated with traditional simple regression analysis. Moreover, causal path analysis allows the *simultaneous estimation* of the regression parameters.

In the next section we present the results of the estimated models for the case of the North and the South of Italy. Before dealing with the results, we discuss the variables used in our Causal Path Analysis.

9.4. Choice of the variables

As we explained from a conceptual point of view in the previous sections, there are some crucial micro-conditions for the exploitation of network externalities. In the framework of our analysis five latent variables have been chosen, representing the micro-conditions for network externality exploitation identified on both the demand and the supply side. On the demand side we expect to find the following micro-conditions:

a) *organisational flexibility* of the firm. The higher the connectivity, the greater the capacity of a firm to be flexible in its organisational structure, since learning and imitative processes are set in motion which may help to overcome the rigidity in implementing organisational changes. Moreover, the greater the organisational flexibility of a firm, the greater the capacity of exploiting network externality on the production side;

b) *entrepreneurship* of a firm. The entrepreneurship of a firm is a rather important factor in exploiting network externality. In particular, we expect that the greater the connectivity of a firm, the greater the entrepreneurship, since a greater experience in the adoption and use of these technologies may stimulate the entrepreneurial capacity necessary to achieve an innovative use of these technologies. A greater degree of entrepreneurship guaranteeing a greater innovative capacity inevitably leads to a greater exploitation of production network externality.

On the supply side, on the contrary, we identify:

c) *help from the supply side,* in terms of technical and organisational assistance given to adopters. From this variable we expect that the greater the connectivity of a firm, the greater the help from the supply side, and consequently a greater exploitation of production network externalities;
d) *importance of the technology for the business.* An extremely important condition is that these technologies are seen as crucial elements for solving corporate business problems. The greater the connectivity, the greater the importance of these technologies, since little by little they become strategic to the day-to-day routine of the firm. The strategic use of these technologies leads, therefore, to the exploitation of production network externalities;
e) *number of firms already connected to the network.* The use of these technologies is guaranteed only if a critical mass is achieved. Thus, we expect that the greater the connectivity, the greater the number of firms replying that "the number of already existing subscribers" is the chief reason for entering the network, since the critical mass has been achieved. The greater the number of firms replying that "the number of other firms in the network" is one of the most important reasons for adoption (i.e. if a critical mass is achieved), the greater the exploitation of production network externalities.

Variables have been specified for each of these conditions, in order to test our model. In particular, we have defined:

a) a variable measuring organisational flexibility. This variable has been defined as the number of functions where organisational changes have occurred in relation to the introduction of telecommunications technology;
b) a variable reflecting the number of innovations (process, product, managerial, or other) which have taken place in the firm has been introduced as a proxy for the entrepreneurial capability existing in a firm;
c) a dummy variable measuring the "help from the supply side", interpreted as a 0-1 variable, assuming a value 1 when firms replied that the

demonstration and promotion centres have played an important role in the diffusion process of these technologies, and 0 otherwise;
d) a dummy variable measuring the importance of these technologies in the business, assuming value a 1 when firms replied that these technologies have been strategic for their business, and 0 otherwise;
e) a dummy variable dealing with the number of firms already existing in the network, and built as a 0-1 variable, assuming a value 1 when firms replied that the most important reason for adoption was the number of already existing subscribers, and 0 otherwise.

Regional differences are expected in the existence of the micro-conditions, which enable firms to exploit production network externality. This expectation stems from the results we obtained from the previous analysis in Chapter 8, where we measured production network externalities. The interesting result of the correlation analysis between the connectivity and the performance index showed that strong regional differences exist in the capacity to exploit production network externality. In particular, the results showed a clear inability of Southern firms to exploit production network externality in comparison with the North. For this reason, we expect that while *in the North these micro-conditions exist* and confirm our conceptual model, a different situation is expected to characterise the South. As we will see hereafter, the results are interesting, for both the North and the South of Italy.

9.5. Results of the estimated "causal path analysis" model for the North of Italy

In this section we concentrate on the results obtained by our estimated causal path analysis model for the North of Italy. For this geographical area, we expect to find a confirmation of our conceptual causal path analysis model. In fact, since a correlation exists between the connectivity and the performance index, our conditions explaining this correlation have to be tested to check that they are statistically significant and parameters have to assume the right sign in order to confirm our conceptual causal path analysis model empirically.

The results obtained are satisfactory. Figure 9.3 shows both the values of the estimated parameters and the T-Student test results for the estimated parameters (presented in brackets). In interpreting the results of both the North and the South of Italy, it should be kept in mind that all variables have been standardised with unit variance and mean zero, in order to be able to compare the relevance of individual parameters.

Before interpreting these results in more detail, some conclusions can already be drawn:

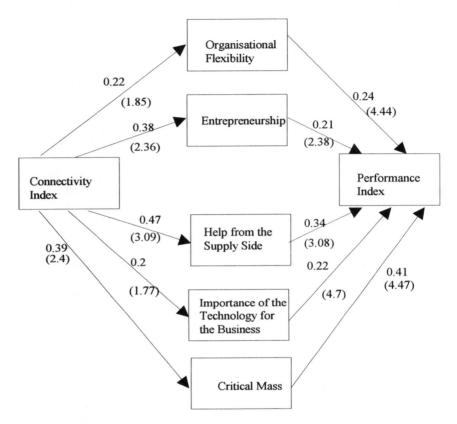

Figure 9.3. Estimated path analysis model for the North of Italy

a) almost all parameters are statistically significant, having T-Student values over 2;
b) all estimated relations have the expected sign;
c) the model itself fits very well with the given variables (P-value = 0.01).

Our conceptual model thus appears to be confirmed. In particular, the connectivity index explains to a large extent the importance of the help from the supply side, the critical mass and the degree of entrepreneurship. These three variables are all statistically significant, and have high estimated parameter values. On the other hand, the most significant variables explaining the performance index have turned out to be the critical mass, followed by the help from the supply side. Also the signs of the parameters are quite satisfactory. The results show that:

a) the greater the connectivity, the greater the organisational flexibility of firms, resulting from the exploitation of learning and imitative processes that are able to overcome the rigidity of organisational corporate structure. Moreover, the greater the organisational flexibility, the higher the impact on the performance of the firm. In this case the parameter is highly statistically significant (4.44);
b) the greater the connectivity, the greater the entrepreneurship of a firm. This result is simply explained by the greater experience which can be built up from previous adoptions and from a long use of these technologies. In this case the parameter assumes a high value (0.38). The model guarantees also that the greater the degree of entrepreneurship, the greater the performance, as we expected;
c) the greater the connectivity, the greater the help from the supply side. The role of the supply side to overcome the technical, managerial and organisational problems related to the adoption of these new telecommunications technologies is of strategic importance. The weight of this linkage is quite heavy, since the parameter has a high value (0.47). As expected, the greater is the help from the supply side to adopting firms, the greater the impact on the production function. The weight of this arrow in the model is quite important, assuming a value of 0.34;
d) the greater the connectivity, the greater the importance of these technologies for the business needs of adopters. This result witnesses the fact that these technologies become strategic for business purposes, after they are in full operation. Even more strategic appears to be the relationship between the importance of these technologies for the business and the performance of firms. This relationship emerges as highly significant (4.7);
e) the greater the connectivity, the greater the number of firms interpreting network externalities as one of the main reasons for adoption. This result is in line with the expectations, since the intense use of these technologies is guaranteed only if a critical mass is achieved. In this case the parameter assumes a very high value (0.39), witnessing the strategic importance of the linkage between these two variables. Moreover, the higher the number of firms considering network externalities as the main reason for adoption, the greater the performance of firms, with a very high statistical significance of the parameter (4.47).

An interesting conclusion which can be drawn from these results is that *the micro-conditions allowing firms to exploit network externalities are present in Northern Italy*. Our analysis would be even more interesting if this were not the case for the South of Italy. This is the subject matter of the next section.

9.6. Results of the estimated causal path analysis model for the South of Italy

A different result is indeed achieved for the South of Italy. As Figure 9.4 shows, the results of the same conceptual model are quite different and in actual fact are less satisfactory, as expected. At first glance one can immediately make a general remark that the model does not fit with the empirical estimations. Most estimated parameters, apart from two of them, are not statistically significant, showing T-Student values below the critical value of 2^5.

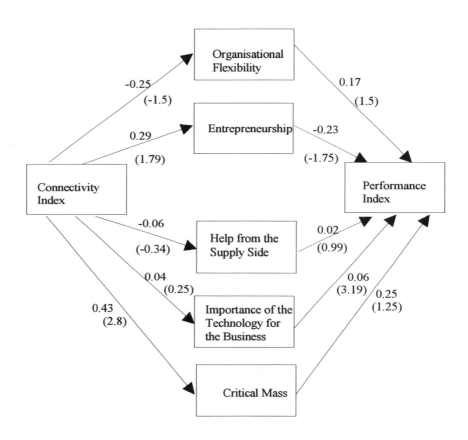

Figure 9.4. Estimated path analysis model for the South of Italy

In any other circumstances this result would have been interpreted as a negative result, destroying from an empirical point of view the conceptual framework underpinning these results. On the contrary, in our case these results support our expectations. The results reflect, in fact, a consistency with what has already been achieved with the empirical analysis in Chapter 8. The main result of that chapter in the case of the South of Italy was that no correlation exists between the connectivity and the performance index. In the present chapter it would have been somewhat surprising to find out that even if firms had shown no capacity for network externality exploitation, the micro-conditions explaining the capacity of firms to exploit production network externalities were to be present in the area. Thus, the results obtained are perfectly in line with the logic of our investigation. A very small number of variables are statistically significant and, moreover, have a very low correlation with the performance and connectivity index.

A general result in this respect is that *micro-conditions to exploit production network externalities seem not to be present in Southern Italy*. We can even argue that with this analysis we have been able to identify the barriers and bottlenecks existing in the exploitation of production network externalities and which hinder the achievement of better economic and spatial performance via the use of advanced telecommunications technologies.

Our conceptual and empirical frameworks thus provide an explanation for the non-existence of a correlation between the connectivity and the performance index. In the South of Italy, the *micro-conditions for the exploitation of network externalities are apparently not present*.

As one may easily see, these results have policy implications which serve as a guide to the right policies which should be introduced as adoption incentives, as we will underline in Chapter 10.

9.7. Conclusions

This chapter has developed an analysis of the micro-conditions for network externality exploitation. The aim of the exercise was to analyse in more detail the relationship between the connectivity and the performance index by defining the conditions under which network externalities are exploited.

In order to achieve this goal, a conceptual causal path analysis framework has been developed. In particular, some conditions have been emphasised which must be present if firms want to exploit production network externalities. These conditions exist on both the demand and the supply side. On the demand side, we refer to the capacity of firms to be flexible in their organisational structure once they adopt these technologies. In fact, these technologies are complex technologies which impact on the transactional structure of the firm, obliging firms to develop organisational changes. Moreover, on the demand side, we expect firms to be able to have a large degree of entrepreneurship, since the adoption of these technologies, with as all kinds of innovation processes, requires a high degree of managerial flexibility and risk strategy.

On the supply side, however, other conditions have to be present, which facilitate the exploitation of production network externalities. In particular, a certain level of adopters has to be achieved in order to generate self-sustained development processes of these technologies. Moreover, usually the supply has to support the demand in the difficult process of adoption, by providing technical and organisational know-how to users. Finally, an indispensable element is a good marketing policy of the supply, underlining the strategic importance of these technologies for business purposes.

The causal path analysis has been tested empirically in the present chapter, in order to confirm whether our third group of hypotheses dealing with the conditions for production network externality exploitation is empirically confirmed.

The estimated causal path analysis model has confirmed our hypotheses, since it has demonstrated that production network externalities are exploited if:

a) a critical mass is achieved;
b) an innovative use of these technologies is achieved;
c) support from the supply side is achieved;
d) a certain degree of entrepreneurship exists, allowing the achievement of competitive advantage via these technologies;
e) a clear importance of these technologies for business purposes is envisaged by users.

These conditions are verified for the case of advanced regions. On the other hand, in the South of Italy the conditions for network externality exploitation are apparently not present. These results could strongly influence future intervention policies devoted to the implementation of advanced telecommunications technologies designed to achieve better economic performance, or even, as in the case of the South of Italy, with a view to decreasing regional disparities. These normative implications will be highlighted in the next chapter, where also some future research directions are pinpointed.

Notes

1. For an in-depth description of causal path analysis, see among others Bollen, 1989; Fuller, 1987; Loehlin ,1987.

2. For an application of causal path analysis, see, among others, Stern and Krakover, 1993.
3. For an example of a causal path analysis Model using another statistical method, i.e. the partial least square method, see Davelaar and Nijkamp, 1992.
4. For an in-depth description of generalised least squares method, see the SAS Manual, 1989.
5. The exercise was run again with the introduction of only the statistically significant variables, i.e. entrepreneurship, critical mass and the importance of the technology for the business, but the results have shown again that the model does not fit well. T-Student values have shown a general lack of statistical significance of the variables introduced.

10 Conclusions and policy implications

10.1. In retrospect

It is widely recognised that advanced telecommunications networks and services are increasingly becoming the strategic weapons for business success in the near future. These new technologies will be instrumental in reshaping the characteristics upon which the competitiveness of firms and comparative advantage of regions will increasingly depend. This idea has been widely accepted and even applied at a normative level within the EC in order to reduce regional imbalances. Innovative policies and especially infrastructure policies aimed at developing advanced telecommunications infrastructure and services, such as the STAR programme, are all initiated with the aim of decreasing economic disparities among regions and achieving "cohesion". This term refers to two concepts at the same time: i) inter-regional convergence of incomes and of development potentials, and ii) increasing co-operation and integration of the regional economies (Camagni, 1993).

Among different intervention policies, the Commission has placed considerable emphasis on the implementation of advanced telecommunications infrastructures and services, with the aim of endowing less developed regions with the most competitive economic tools of the 1990s.

At the same time, the role that advanced telecommunications networks and services are increasingly playing in modern economies has renewed the interest in those mechanisms associated with the diffusion of these technologies, and in their effects within different economic areas. This study provides an insight into this research field, by merging two different areas of study together. In particular, this study is part of the conceptual approach which defines telecommunications as the driving forces for economic deve-lopment and growth, but it goes a step further by assuming that the advantages firms (and regions) gain from these technologies is not just related to the technological changes that this sector is undergoing. The advantages which may be generated by these technologies on the performance of firms also stem from the most important and peculiar characteristic of these technologies as interrelated technologies: the utility and productivity functions are dependent on the number of already existing subscribers. This peculiarity has been studied a lot in the literature in relation to the utility function. This concept is well known as *"network externality"*. Since Rohlf's paper (1974), the concept of interrelation among the utility functions of telecommunications users, and thus of (consumption) network externality, has been widely analysed and discussed. This study goes a step further by claiming that productivity functions are also dependent on the number of subscribers. In fact, a vast survey of the literature has brought to light the lack of any previous effort devoted to the study of network externalities and their effects on the production function.

In this study we claim that *if network externalities are one of the most important reasons for entering the network, better economic performance is the effect they produce*. On the basis of this observation, a conceptual framework has been built, where the performance of firms was highly influenced by the existence of (production) network externality. This study provides a conceptual and empirical answer to the following research issues:

a) *what is the relative importance of network externalities for firms in their decision to join a network*;
b) *whether network externalities have an effect on the performance of firms and regions*;
c) *which micro- and macro-conditions facilitate or hamper the exploitation of production network externalities*.

If our conceptual framework is empirically valid, policy implications arise, since the accessibility to advanced telecommunications network may represent a way of internalising externality mechanisms. In this chapter some policy guidelines are highlighted (Section 10.3). In particular, our study leads to the conclusion that the mere accessibility to networks and services does not allow the internalisation of externality mechanisms and that other strategic policy elements have to be realised in order to restore market equilibrium forces.

As far as the *spatial dimension* is concerned, an extensive survey of the literature on regional aspects of technological change has served to bring to the fore the general idea that all innovations start in the centre, and consequently diffuse over space, despite the unique features of each technology. However, instead of this uniform distribution process, the characteristics of each technology may influence the particular spatial diffusion patterns, and may require different seedbed conditions in order to develop. For this reason, our analysis draws attention to the intrinsic characteristics of these technologies, which may influence spatial development patterns. As we will see hereafter, the results achieved are quite satisfactory (Section 10.2) and are of strategic importance for normative implications (Section 10.3).

10.2. Results achieved

In the empirical part of the study an attempt was made to prove the empirical validity of the framework presented in the theoretical part, by means of testing the hypotheses derived from our theoretical reflections.

The empirical analysis has been run on the basis of a database collected with an in-depth interview methodology for a sample of firms located in both the North and the South of Italy. The contrasting economic development of the two areas has significantly contributed to emphasise the "regional dimension" in the analysis, since all hypotheses could be tested in different economic environments. In particular, for the South of Italy, firms were chosen from among those participating in the so-called STAR (Special Telecommunications Action for Regional Development) programme run by the EC with the aim of developing advanced telecommunications networks and services in less favoured regions of the Community. An evaluation of the STAR effects on the competitiveness of local firms, and thus on regional competitiveness has been run, and some interesting results stem from the analysis. These results are extremely useful to emphasise some *strategic lessons* for future intervention policies.

All three research issues mentioned in Section 10.1 have been analysed at an empirical level and very interesting results have emerged, which confirm our conceptual framework and expectations.

Concerning the *first research issue*, i.e. whether network externalities are one of the main reasons for joining the network, the results partially support this idea. In this case, both a descriptive and an interpretative analysis were run with the aim of testing the hypotheses linked to our first research issue, and in both cases the replies have been positive. The descriptive analysis, based on contingency tables, supports our hypotheses. In fact, the results show that in the first phases of the diffusion process of telecommunications technologies the main reason for having adopted is not the number of already existing subscribers, while the main reason for non-adoption is related to the low number of people connected to the firm via the network. In fact, at the firm level, the major problem which hampers a more intensive use of these technologies is the low number of subscribers. The opposite results hold in the case of a more advanced adoption process.

A relationship exists between the "network externality" variables and the regional dimension. In other words, the most important reasons for joining the network are different at the regional level. In the North, a much more economically developed area where these technologies were adopted some years ago, the main reasons for future adoption are not related to the number of already existing subscribers, since the number of subscribers is large enough to be ignored when choosing a new network with which to be connected. However, a different reality characterises the South of Italy, where the main reasons for future adoption are strongly related to the number of future subscribers. This result shows the fact that in the first stages of diffusion processes (as is the case in the South of Italy) one of the main reasons for future adoption is the number of already existing subscribers.

The interpretative analysis of the determinants of the willingness to make future adoptions, run on the basis of logit model estimates, leads to slightly less satisfactory results. In fact, the estimated logit models for the regional analysis shows that the number of already existing subscribers is not an important explanatory variable for the willingness to adopt, while price incentives and support from the supply side have more explanatory value.

As far as the *second research issue* is concerned, important conclusions may be drawn, which support the hypotheses deduced from our conceptual framework. In particular, the analysis was run in order to test whether network externalities have an effect on the performance of firms and regions. In other words, an attempt was made to test production network externalities. If positive network externalities play a role on the performance of firms, a positive relationship between a connectivity index (representing the level of physical connectivity of the firm to these networks and services), and a performance index is expected. The results obtained by running a simple correlation analysis are quite interesting. A positive correlation exists between the connectivity and the performance index in the case of the North of Italy, while for Southern Italy the correlation does not exist. Once again the spatial dimension has a strategic role in the analysis.

In addition, an analysis of the structural characteristics of firms in our sample was run, with the aim of clustering them for their structural characteristics which could explain their different capacity to exploit network externalities. Since no correlation exists in the South, the identification of other structural characteristics of the firms on the basis of their capacities to exploit network externalities leads of course to no specific results. In the case of the North of Italy, on the contrary, 6 distinct clusters of firms emerge which explain the different structural features of firms having the same capacities to exploit network externalities. In particular, those firms which may be labelled "*winners*" for their greater capacity to exploit network externalities, are also characterised by organisational flexibility and innovative capacity, both of these qualities being very much higher than the average. On the contrary, firms which may be labelled "*losers*" for their inability to exploit network externality are also those which have very low levels of flexibility and innovative capability. Within these two categories, a range of three other groups was determined, with a different combination of structural variables.

The definition of these structural characteristics accompanying the exploitation of network externalities suggests that some *conditions* exist in order to exploit network externalities. The existence of these conditions is extremely important; in fact, it would be misleading to think about a simple linear relationship between the mere adoption and better economic performance. This relationship depends on a complex set of variables, whose existence determines the exploitation of network externalities.

On the basis of this idea, the *third research issue* was empirically analysed with the aim of: a) understanding the relationship between the connectivity and the performance index, by emphasising the elements which allow the relationship to take place; b) formulating normative implications for future intervention policies. The empirical analysis was run on the basis of a conceptual path analysis framework. In this framework crucial micro-conditions have been identified, which facilitate the exploitation of production network externality. In particular, these conditions, widely mentioned in the literature, which influence the diffusion processes of telecommunications technologies, have been identified on both the demand and the supply side. On the *demand side*, they are related to the organisational flexibility of the firms in adopting these technologies. These technologies are in fact complex technologies and impact on the transactional structure of a firm, changing the way firms interact and organise their internal and external relationships. For this reason, the adoption and especially the intense use of these technologies require a high degree of organisational and technical know-how in order to be able to exploit their high potentialities. On the demand side, the entrepreneurial capacity of a firm is naturally of strategic importance. All organisational changes required to adopt these technologies are accepted and implemented when a certain degree of entrepreneurship and risk strategy is present in the firm. The presence of these corporate characteristics guarantees the acceptance and full exploitation of these technologies on the demand side.

As far as the *supply side* is concerned, other strategic elements have to be present in order to ensure the satisfactory exploitation of production network externalities. First of all, the achievement of a certain level of subscribers allows self-sustained mechanisms to be set in motion which support the adoption processes. Moreover, managerial, organisational and technical help from the supply side can greatly facilitate the adoption mechanisms. Finally, a good marketing policy should focus not only on the technical potentialities of these technologies but also on the importance that these advanced telecommunications networks and services can have for the business concerned. If this strategic role is emphasised, the possibilities of exploiting network externalities are higher.

All these conditions have been expressed in a causal path model and its empirical validity has been estimated with the use of the Generalised Least Squares Method. This method, like other statistical methods such as the partial least squares method and the Lisrel method, makes it possible to run simultaneous regression estimates of parameters belonging to an equation system reflecting the causal path among variables. The estimated parameters and their statistical significance for the North confirmed empirically what our conceptual framework of causal path analysis was suggesting. As expected, the micro-conditions for the exploitation of network externalities are present in the North and reflect the positive sign of the relationship between the connectivity and the performance index. This is not the case for the South of Italy, where the estimated parameters appear to be insignificant. The results are perfectly in line with the previous analysis (Chapter 8), which showed no correlation between the connectivity and the performance index.

From this analysis, some extremely interesting policy implications emerge in the field of infrastructure policies which aim at increasing economic growth, and in particular at decreasing regional economic disparities. This is the subject matter of the next section.

10.3. Policy implications

The empirical analysis run in this study contains some *policy implications* regarding the effects these technologies have in reducing inter-regional disparities. In particular, a first crucial result is that mere accessibility to advanced telecommunications infrastructures and services does not necessarily lead to a better corporate and regional performance. The results in the South of Italy are representative in this respect, since the empirical analysis shows no correlation between greater intense use of these technologies and better industrial and regional performance. From such results, one may be inclined to draw a negative conclusion about the effects that the STAR intervention programme run by the EC has generated on the performance of local firms. However, such a conclusion is far too pessimistic and could be highly criticised, since:

a) evaluations of regional impact require a long-term perspective. Regional effects take place in the long-run, and only a long-term evaluation can really test the benefits of the programme. Our analysis took place only a few months after the end of the programme, when all the spin-off and spillover effects at regional level may still not have generated all their effects;
b) in the short run, an evaluator can only be concerned with the *potential* that these technologies have to influence regional performance. In this respect, this study is useful, since it emphasises *bottlenecks* and *barriers* which hamper the exploitation of production network externalities by firms. The conditions put forward by our conceptual framework and tested in our empirical analysis in Chapter 9 in order to overcome these bottlenecks and

barriers suggest some policy recommendations. These recommendations should be taken into consideration when developing an infrastructure intervention policy, like STAR, which aims to decrease regional disparities.

The first crucial precondition to assure a sustainable adoption of these technologies (and thus greater real connectivity) is the achievement of a *critical mass* of adopters especially for interrelated services, such as electronic mail. The user value of these technologies is in fact related to the number of already existing subscribers, since the attraction for a new potential subscriber to join the network is the possibility of being linked to a great number of subscribers. The existence of a certain number of subscribers is a necessary condition to stimulate a *cumulative self-sustained mechanism*. The number of subscribers at which this self-sustained mechanism takes place varies according to the sector to which firms belong. Some sectors are more inclined to accept the risk of a low number of subscribers, since telecommunications technologies are in any case extremely important for their business, as it was also pointed out in our empirical analysis (see Chapter 7). A policy recommendation underlining the importance of a number of subscribers may appear rather too general, since the critical mass level is extremely difficult to be measured at an empirical level (see also Allen, 1990a). However, it is our opinion that the general suggestion on the importance of the number of subscribers in a network has to be stressed in telecommunications development policy guidelines, since many programmes on telecommunications implementation (at both European and National levels) have demonstrated to underestimate a so strategic development tool, probably because of the high level of financial investments required to exploit it.

Another critical factor for the achievement of a great number of adoptions is the *clear identification of the way these technologies may be useful for business purposes*. These technologies are by no means simple technologies and require a certain degree of organisational changes in order to become useful to business needs. This process requires a strong organisational effort on the part of the subscriber, who most of the time needs organisational support from specialised telecommunications experts. The STAR programme has indeed taken this aspect into account. Some measures were devoted to the implementation of "demonstration and promotion actions", through specialised centres whose task was to develop a "telematic culture" among potential users (the so-called 4.2 measures). However, as the results underline, the crucial element in this process is not only the "technical aspect", but rather the "*organisational aspect*", because of the complexity of integrating these technologies within the established business structure. From the results obtained, it seems that these centres promoted these services underlining in most cases their technical capacities, and assuring technical support to firms. They did not, however, deal sufficiently well with organisational problems and changes which have to be borne in order to exploit these technologies in the best way.

The relatively low intensity of use of these technologies is explained in our empirical analysis of the reasons for non-adoption (see Chapter 7) by the inability of subscribers to use these technologies for corporate needs. In this respect the following question is very important: Are these technologies of any real interest to small- and medium-sized firms, or is it a matter of how these technologies have been promoted which has failed to demonstrate the potentialities linked to the business? Our inclination is to reply positively to the second possibility. It is not a matter of the non-utility of these technologies to the business needs of small- and medium-sized enterprises in general; it is much more a matter of finding the right incentive policies to promote them, which is not a simple task. It is not a simple task in developed areas, where frictions and obstacles to the organisational changes exist, which all hamper this process. It is an even more complex task in less developed regions, where the economic environment is not very conducive to generating *entrepreneurship* and *economic stimuli* to changes.

The economic environment of the South, as is generally the case in less developed regions, is weak in terms of traditional crucial factors which stimulate the adoption of technological change. In fact, in the South the economic environment is characterised by: a) a very high risk aversion by potential adopters, b) very limited competitive market forces, c) a stable division of market shares. All these features define a very difficult environment where self-sustained adoption mechanisms of new technologies risk finding no stable ground in the short term and where for these reasons the usual incentive policies have to be extremely strong especially in the first phases of the diffusion processes. The STAR programme heavily subsidised networks and services during these early years (to the extent that these technologies were free of charge for the pioneering users). This incentive policy has turned out to be extremely useful, as our analysis has clearly demonstrated (see Chapter 7). However, there has been no adequate policy underlining the potentialities of these technologies in terms of business needs solutions or even in terms of new market niches to be exploited. The provision of these complex technologies has to cover especially the *organisational aspects*, rather than the simple technical details. Moreover, a successful provision requires the identification of the ways in which these technologies can be exploited to solve *business needs* or to achieve new *business goals*. The organisational aspect and the business idea aspect have not been regarded as strategic factors for adoption; neglect of these two aspects could create serious bottlenecks in the adoption process and hamper the exploitation of (production) network externalities by firms and, via multiplicative effects, by regions.

All these remarks lead to the identification of some *policy guidelines* for future intervention policies which have to be taken into account to achieve successful results.

First of all, from all that has been said above, it is possible to claim that the promotion policies of these technologies have to be based on a *bridging mechanism between demand and supply*, i.e. they have to be able to link business needs, or even potential business needs, to the existing technological potentialities. Suppliers should be able to provide not only the physical infrastructure and technical support, but also provide the business angle on how to exploit these technologies, on the basis of business needs of potential users. One way of dealing with this problem is to customise these networks and services as much as possible to the personal needs of potential users.

A second lesson from the STAR programme is more related to the "spatial circumstances" in which these technologies have to be promoted. A crucial resource for the development of these technologies is *entrepreneurship*. In other words, the presence of risk aversion and of non-competitive market structures discourages adoption processes of these technologies, since no market force exists under those conditions which can stimulate firms to bear the organisational, managerial and financial costs necessary for a successful adoption. Thus, local entrepreneurship turns out to be a strategic element for successful adoptions. If this is the case, innovative policies have to take this aspect into consideration, choosing, among other factors, local areas where *entrepreneurial capabilities* are available. Thus, instead of supplying these technologies to all areas in less developed regions, as has been the case with the STAR programme in Italy, innovative policies should rather be focused on the most dynamic areas in terms of both technical and entrepreneurial capabilities. These areas, these "*local milieux*", represent the most efficient and dynamic areas, where a technological policy could lead in the long run to good results in terms of network externality exploitation. This modern view of intervention policies suggests that technology policy, when implemented without a territorial perspective, either results in a cumulative process of spatial concentration of technological development, or lazily remains inefficient, as it overlooks the adoption problems of small- and medium-sized enterprises[1].

Another important aspect in the technology policy for regional development is its capacity to overcome *local economic constraints*. Telecommunications technologies are seen by the users as a way to achieve information and know-how which are not present locally, and especially which are typical of advanced economic areas. For this reason, the STAR programme is seen as crucial for shrinking the (physical and economic) distance between backward and more advanced areas in the Community. However, the way in which the programme has been run and managed, guaranteeing an advanced link in backward areas, has lost part of its attractiveness. The creation of *a club of poor regions* does not seem to be the right policy guideline for ensuring local sustainable economic development. In the Italian case, the implementation of advanced networks and services in the South, with no direct link to the North of Italy, has acted as a disincentive for adoption.

The impact of STAR on regional development is geared to overcoming these bottlenecks and barriers which hamper the full exploitation of these technologies. The capacities of future intervention policies aiming to solve these problems will determine the extent to which the implementation of these technologies will influence regional performance in the future. For this reason, it is of vital importance to assure continuity in the provision of these technologies, via the launch of a second intervention programme. This would have two aims: a) to overcome the present bottlenecks and barriers in the adoption processes, taking into account the "lessons" learnt by the first phase of the programme, b) to achieve the expected positive effects on the regional performance, at present hampered by both the low level of adoption and of use.

10.4. New research areas

This study has made an in-depth examination of the effects that network externalities have on the productivity of firms and regions. This is rather a new field of research and this study is an attempt to extend the frontiers of knowledge in this area. However, there is still some scope for further research efforts. In particular, we identify four major areas:

a) an extension of the present study;
b) an analysis of the effects that other kinds of "network externalities" may have on the level of performance of firms;
c) an analysis of other kinds of "networks", where the concept of network externality may be applied;
d) other aspects related to the existence of network externality;
e) an analysis of the possible ways to overcome this kind of externalities.

These areas of research studies are analysed in detail hereafter.

10.4.1. An extension of the present study

A first field of scientific interest for future research is a *dynamic analysis* of the "economic and spatial symbiosis" framework. The present work is a *static interpretation* of the effects network externalities generate on the performance of firms. The first immediate follow-up of the study could be a *dynamic interpretation*, at both a conceptual and an empirical level. At the conceptual level, it would be rather interesting to analyse how firms and regions change their capacity to exploit network externalities over time. Our work suggests that when the conditions for the exploitation of network externalities are present, it is possible to foresee a linkage between the degree of physical connectivity of firms and corporate and regional performance. A dynamic analysis should focus on how these conditions evolve in an area over time. This would imply the collection of empirical data in time series, in order to see if and when the capacity of firms to exploit network externalities changes over time. The present study has not considered the *expectations* that potential entrants may have concerning the future subscriber base and which may influence the decision to join a network. This aspect may explain why our analysis on the willingness of adoption based on the actual subscriber-base did not lead to very satisfactory results. The expectations concerning the future number of subscribers undoubtedly deserves attention in future research activities.

Another interesting and stimulating area of work which remains open in this study is the interpretation of the *spatial aspects* in a more direct way. This analysis has treated the spatial dimension via the location of firms, and indirectly assumed that the effects generated by these technologies on the production function of firms would have been transmitted to the aggregate production function via multiplicative effects. A follow-up of this analysis could emphasise regional aspects more directly, by identifying particular macro-conditions which can influence the exploitation of network externalities.

10.4.2. Other kinds of network externalities

Our model is based on the assumption that *technical* network externalities play a role on the performance of firms. It would be extremely interesting to develop a model where *pecuniary* network externalities play a role in the profit function of firms, and see whether this kind of network externalities also exist at an empirical level. In this case, the assumption should be the link between the use of telecommunications technologies and the profit function of these firms and some propositions could be derived on the basis of a profit maximising model approach. The question is whether one would expect to find similar results to those for the case of technical network externalities.

10.4.3. Other kinds of networks

Another potentially fruitful research area is the possible application of the concept of network externality to other kinds of (immaterial) networks, for example "firms' networks". In particular, the concept of network externality could be extremely useful to explain the dynamics of the so-called local districts, defined as territorial areas with a concentration of vertically integrated firms. In these local districts, the concept of network externality may be associated to the advantages that these firms obtain by being located in the area, and by being a member of this (immaterial) network of firms. The advantages, which stem mainly from the location of firms, may be labelled "local network externalities", and may be defined as (Camagni, 1991c):

a) a greater capacity of the firm to collect and gather information, through informal channels of information between firms operating in the same markets;
b) a greater capacity to monitor the direction of the market;
c) a more specialised and skilled labour force, stemming from collective learning processes;
d) a collective process of definition of managerial styles and decision routines, achieved through managerial labour mobility, imitative decisions, cooperative decision-making through local industrialists' associations complementary innovation processes;
e) a greater capacity to achieve collective decisions, through interpersonal linkages.

These advantages increase when a new firm enters the (immaterial) network, i.e. when a new firm locates in the area. However, at the same time costs also increase related to the location of a new firm, since the labour force becomes more scarce (or more expensive), and the local market has to be divided between a greater number of firms[2]. While the marginal benefits of being located in a local district increase, marginal costs also increase and the game is worthwhile until marginal costs are greater than marginal benefits. In this situation, the location in the district is no longer an interesting choice for a firm, since negative network externalities start to play a role.

It would be rather interesting to extend this analysis in both theoretical and empirical terms, on the basis of the concept of network externality. This could certainly represent a way to interpret the dynamics of these local economies.

10.4.4. *Other aspects related to network externalities*

As we mentioned in Chapter 2, other aspects are related to the existence of network externality, which do not belong directly to the sphere of the impact that network externalities have on the production function. These aspects include standardisation, interconnectivity of different networks and tariff structure. One extremely interesting analysis could be to link these aspects to our framework by identifying, for example, the effects that greater interconnectivity among networks could generate in our model. In dynamic terms, the marginal benefits of being networked follow a S-shaped curve (see

Chapter 4). A plausible question would be to what extent interconnectivity could act on production network externality exploitation. The same kind of question could be related to the achievement of similar technical standards, by trying to answer the question of to what extent compatible standards could produce an advantage for firms and regions in order to exploit network externalities.

10.4.5. *Different ways to overcome this kind of externalities in the market*

Another interesting research area which is not explored by this study relates to the ways in which this kind of externalities may be overcome in a market. Three are the possible causes for their persistent presence in the telecommunications sector:

a) transaction costs for internalisation of the externality mechanisms are too high;
b) network externalities are the result of a short term phenomenon provoked by too low prices;
c) absence of property rights which can regulate the market.

All these three explanations are plausible in the telecommunications industry. It is exactly on these "distortion mechanisms" that policy makers should act in order to internalise the externalities into the price system. On which of the three causes it is better to act remains an open subject. Our opinion is that the answer highly depends on the institutional system which regulates the telecommunications system and thus, it is more a national solution rather than a global rule which can be applied commonly at a European or International scale.

The unexplored areas related to the concept of network externality and its impact on the performance of firms and regions are still manifold. While this study does provide a new contribution it is only a first attempt to conceptualise the issue and to provide empirical evidence to support the suggested conceptual framework. The subject of network externalities offers a rich seam of research opportunities so far largely unmined.

Notes

1. See Camagni (1993) for a wider interpretation of innovation policies at the local level.
2. For an extended definition of this concept see also Antonelli, 1993.

Annex 1:
Mathematical specifications of the model

In this annex we present the mathematical solution of the equation systems which allowed us to obtain the results presented in Chapter 5.

We begin with *the case of two symmetric firms* in a market. From the minimisation of the langrangian function for N one obtains:

$$N_1 = \frac{\varepsilon_{11} \delta_1 K_1}{\alpha_1} \cdot \frac{1}{\frac{c_{I_1}}{c_{K_1}} + c_{n_1} \varepsilon_{11}} \tag{1}$$

The specific shape of the Cobb-Douglas production function and the minimisation of the lagrangian function for K and L imply:

$$q_1^* = \eta_1 K_1^{\alpha_1 + \beta_1} \left(\frac{c_{K_1} \alpha_1}{c_{L_1} \beta_1} \right)^{\beta_1} N^{\delta_1} = \eta_1 K_1^{\alpha_1 + \beta_1} A_1^{\beta_1} N_1^{\delta_1} \tag{2}$$

where $A = \dfrac{c_{K_1} \alpha_1}{c_{L_1} \beta_1}$ \hfill (3)

From equation 2, one obtains the value of K:

$$K_1 = \left(\frac{q_1^*}{\eta_1 A_1^{\beta_1}}\right)^{\frac{1}{\alpha_1+\beta_1}} \frac{1}{N_1^{\frac{\delta_1}{\alpha_1+\beta_1}}} \qquad (4)$$

If we substitute equation (4) in equation (1) one obtains:

$$N_1^{\frac{\alpha_1+\beta_1+\delta_1}{\alpha_1+\beta_1}} = \frac{\varepsilon_{11}\delta_1}{\alpha_1}\left(\frac{q_1^*}{\eta_1 A_1^{\beta_1}}\right)^{\frac{1}{\alpha_1+\beta_1}} \frac{1}{\frac{c_{I_1}}{c_{K_1}}+c_{n_1}\varepsilon_{11}} \qquad (5)$$

Since $\alpha_1 + \beta_1 + \delta_1 = 1$, one obtains:

$$N_1 = \frac{q_1^*}{\eta_1 A_1^{\beta_1}}\left(\frac{\varepsilon_{11}\delta_1}{\frac{\alpha_1 c_{I_1}}{c_{K_1}}+\alpha_1 c_{n_1}\varepsilon_{11}}\right)^{\alpha_1+\beta_1} \qquad (6)$$

If one substitutes A (defined in equation 3) in equation (6), one has:

$$N_1 = \frac{q_1^*}{\eta_1}\left(\frac{c_{L_1}\beta_1}{c_{K_1}\alpha_1}\right)^{\beta_1}\left(\frac{\delta_1}{\frac{\alpha_1 c_{I_1}}{\varepsilon_{11}c_{K_1}}+\alpha_1 c_{n_1}}\right)^{\alpha_1+\beta_1} \qquad (7)$$

Sustituting equation (7) in the connectivity function $N_1 = \varepsilon_{11}I_1 + \varepsilon_{21}I_2$, one obtains:

$$I_1 = \frac{q_1^*}{\varepsilon_{11}\eta_1}\left(\frac{c_{L_1}\beta_1}{c_{K_1}\alpha_1}\right)^{\beta_1}\left(\frac{\delta_1}{\frac{\alpha_1 c_{I_1}}{\varepsilon_{11} c_{K_1}} + \alpha_1 c_{n_1}}\right)^{\alpha_1+\beta_1} - \frac{\varepsilon_{21}}{\varepsilon_{11}}I_2 \qquad (8)$$

which is the equation expressed in Chapter 5, Section 5.3.1.

In the case of two asymmetric firms in the market, the situation for the follower is the minimisation of its number of contacts, i.e. equation (7) expressed for the case of firm 2:

$$I_2 = \frac{q_2^*}{\varepsilon_{22}\eta_2}\left(\frac{c_{L_2}\beta_2}{c_{K_2}\alpha_2}\right)^{\beta_2}\left(\frac{\delta_2}{\frac{\alpha_2 c_{I_2}}{\varepsilon_{22} c_{K_2}} + \alpha_2 c_{n_2}}\right)^{\alpha_2+\beta_2} - \frac{\varepsilon_{12}}{\varepsilon_{22}}I_1 = B - \frac{\varepsilon_{12}}{\varepsilon_{22}}I_1 \qquad (9)$$

where $B = \dfrac{q_2^*}{\varepsilon_{22}\eta_2}\left(\dfrac{c_{L_2}\beta_2}{c_{K_2}\alpha_2}\right)^{\beta_2}\left(\dfrac{\delta_2}{\frac{\alpha_2 c_{I_2}}{\varepsilon_{22} c_{K_2}} + \alpha_2 c_{n_2}}\right)^{\beta_2}$ \qquad (10)

Substituting equation (9) in the connectivity function $N_1 = \varepsilon_{11}I_1 + \varepsilon_{21}I_2$, one obtains:

$$N_1 = \varepsilon_{11}I_1 + \varepsilon_{21}\left(B - \frac{\varepsilon_{12}}{\varepsilon_{22}}I_1\right) \qquad (11)$$

where $B = \dfrac{q_2^*}{\varepsilon_{22}\eta_2}\left(\dfrac{c_{L_2}\beta_2}{c_{K_2}\alpha_2}\right)^{\beta_2}\left(\dfrac{\delta_2}{\dfrac{\alpha_2 c_{I_2}}{\varepsilon_{22}c_{K_2}} + \alpha_2 c_{n_2}}\right)$

The first partial derivative of equation (11) with respect to the number of contacts is:

$$\frac{\delta N_1}{\delta I_1} = \varepsilon_{11} - \frac{\varepsilon_{21}\varepsilon_{12}}{\varepsilon_{22}} \tag{12}$$

The first order condition implies:

$$\frac{\delta q_1}{\delta N_1}\frac{\delta N_1}{\delta I_1} = \delta_1 \frac{q_1^*}{N_1}\left(\varepsilon_{11} - \frac{\varepsilon_{21}\varepsilon_{12}}{\varepsilon_{22}}\right) = \frac{c_{I_1}}{\lambda} + c_{n_1}\alpha_1 \frac{q_1^*}{K_1}\left(\varepsilon_{11} - \frac{\varepsilon_{21}\varepsilon_{12}}{\varepsilon_{22}}\right) \tag{13}$$

thus N_1 is equal to:

$$N_1 = \left(\varepsilon_{11} - \frac{\varepsilon_{21}\varepsilon_{12}}{\varepsilon_{22}}\right)\frac{\delta_1 K_1}{\alpha_1}\frac{1}{\dfrac{c_{I_1}}{c_{K_1}} + \left(\varepsilon_{11} - \dfrac{\varepsilon_{21}\varepsilon_{12}}{\varepsilon_{22}}\right)c_{n_1}} \tag{14}$$

The shape of the Cobb-Douglas production function and the minimisation of K and L imply:

$$K_1 = \left(\frac{q_1^*}{\eta_1 A_1^{\beta_1}}\right)^{\frac{1}{\alpha_1+\beta_1}}\frac{1}{N_1^{\frac{\delta_1}{\alpha_1+\beta_1}}} \tag{15}$$

where $A = \dfrac{c_{K_1} \alpha_1}{c_{L_1} \beta_1}$ \hfill (16)

If one substitutes equation (15) in equation (14), one has:

$$N_1 = \left(\varepsilon_{11} - \dfrac{\varepsilon_{21}\varepsilon_{12}}{\varepsilon_{22}}\right)\dfrac{\delta_1}{\alpha_1}\left(\dfrac{q_1^*}{\eta_1 A_1^{\beta_1}}\right)^{\frac{1}{\alpha_1+\beta_1}} \dfrac{1}{N_1^{\frac{\delta_1}{\alpha_1+\beta_1}}} \dfrac{1}{\dfrac{c_{I_1}}{c_{K_1}} + \left(\varepsilon_{11} - \dfrac{\varepsilon_{21}\varepsilon_{12}}{\varepsilon_{22}}\right)c_{n_1}} \hfill (17)$$

which also means:

$$N_1^{\frac{\alpha_1+\beta_1+\delta_1}{\alpha_1+\beta_1}} = \left(\varepsilon_{11} - \dfrac{\varepsilon_{21}\varepsilon_{12}}{\varepsilon_{22}}\right)\dfrac{\delta_1}{\alpha_1}\left(\dfrac{q_1^*}{\eta_1 A_1^{\beta_1}}\right)^{\frac{1}{\alpha_1+\beta_1}} \dfrac{1}{\dfrac{c_{I_1}}{c_{K_1}} + \left(\varepsilon_{11} - \dfrac{\varepsilon_{21}\varepsilon_{12}}{\varepsilon_{22}}\right)c_{n_1}} \hfill (18)$$

Since $\alpha_1 + \beta_1 + \delta_1 = 1$, one obtains:

$$N_1 = \left(\dfrac{\left(\varepsilon_{11} - \dfrac{\varepsilon_{21}\varepsilon_{12}}{\varepsilon_{22}}\right)\delta_1}{\dfrac{\alpha_1 c_{I_1}}{c_{K_1}} + \alpha_1\left(\varepsilon_{11} - \dfrac{\varepsilon_{21}\varepsilon_{12}}{\varepsilon_{22}}\right)c_{n_1}}\right)^{\alpha_1+\beta_1} \dfrac{q_1^*}{\eta_1 A_1^{\beta_1}} \hfill (19)$$

where $A = \dfrac{c_{K_1}\alpha_1}{c_{L_1}\beta_1}$ \hfill (20)

If one equals equation (19) with equation (11), one obtains:

$$I_1 = \dfrac{1}{\varepsilon_{11} - \dfrac{\varepsilon_{21}\varepsilon_{12}}{\varepsilon_{22}}} \dfrac{q_1^*}{\eta_1} \left(\dfrac{c_{L_1}\beta_1}{c_{K_1}\alpha_1}\right)^{\beta_1} \left(\dfrac{\delta_1}{\dfrac{\alpha_1 c_{I_1}}{\left(\varepsilon_{11} - \dfrac{\varepsilon_{21}\varepsilon_{12}}{\varepsilon_{22}}\right)c_{K_1}} + \alpha_1 c_{n_1}}\right)^{\alpha_1+\beta_1} - \dfrac{\varepsilon_{21}}{\left(\varepsilon_{11} - \dfrac{\varepsilon_{21}\varepsilon_{12}}{\varepsilon_{22}}\right)} \cdot A$$

where $A = \dfrac{q_2^*}{\varepsilon_{22}\eta_2}\left(\dfrac{c_{L_2}\beta_2}{c_{K_2}\alpha_2}\right)^{\beta_2} \left(\dfrac{\delta_2}{\dfrac{\alpha_2 c_{I_2}}{\varepsilon_{22}c_{I_2}} + \alpha_2 c_{n_2}}\right)^{\alpha_2+\beta_2}$

This is the equation represented in Section 5.3.2. of Chapter 5.

The case of three firms implies the following connectivity function:

$$N_1 = \varepsilon I_{1,j} + \mu I_{j,1} + \tau I_{j,h}$$

The optimal level of $I_{1,j}$ is similar to the case of Cournot. From the minimisation, we obtain:

$$I_{1,j} = \frac{q_1^*}{\varepsilon \eta_1} \left(\frac{c_{L_1} \beta_1}{c_{K_1} \alpha_1} \right)^{\beta_1} \left(\frac{\delta_1}{\frac{\alpha_1 c_{I_1}}{\varepsilon c_{K_1}} + \alpha_1 c_{m_1}} \right)^{\alpha_1 + \beta_1} - \frac{\mu}{\varepsilon} I_{j,1} - \frac{\tau}{\varepsilon} I_{j,k} \qquad (21)$$

where $\varepsilon I_{1,j} = \varepsilon_{1,2} I_{1,2} + \varepsilon_{1,3} I_{1,3}$

In order to find the number of contacts that firm 1 receives from firm 2, and separate them from the number of contacts firm 1 receives from firm 3, one obtains:

$$I_{1,3} = \frac{\varepsilon - \varepsilon_{1,2}}{\varepsilon_{1,3} - \varepsilon_{1,2}} I_{1,j}$$

$$I_{1,2} = \frac{\varepsilon - \varepsilon_{1,3}}{\varepsilon_{1,2} - \varepsilon_{1,3}} I_{1,j}$$

The cases of serial and partial connectivity are two particular cases of the full connectivity. In particular, for the *serial connectivity* the general term j is equal to 2, and k to 3. This leads to the following equation:

$$I_{1,2} = \frac{q_1^*}{\varepsilon \eta_1} \left(\frac{c_{L_1} \beta_1}{c_{K_1} \alpha_1} \right)^{\beta_1} \left(\frac{\delta_1}{\frac{\alpha_1 c_{I_1}}{\varepsilon c_{K_1}} + \alpha_1 c_{m_1}} \right)^{\alpha_1 + \beta_1} - \frac{\varepsilon_{2,1}}{\varepsilon_{1,2}} I_{2,1} - \frac{\tau}{\varepsilon_{1,2}} I_{2,3} \qquad (22)$$

which is the equation presented in Section 5.4.1 of Chapter 5.

In the case of *partial connectivity*, no connections exist between firm 2 and 3. Thus,

$$I_{2,3} = 0$$

This allows us to say that the optimal number of contacts between firm 1 and 2 is:

$$I_{1,j} = \frac{q_1^*}{\varepsilon \eta_1} \left(\frac{c_{L_1} \beta_1}{c_{K_1} \alpha_1} \right)^{\beta_1} \left(\frac{\delta_1}{\frac{\alpha_1 c_{I_1}}{\varepsilon c_{K_1}} + \alpha_1 c_{n_1}} \right)^{\alpha_1 + \beta_1} - \frac{\mu}{\varepsilon} I_{j,k} \qquad (23)$$

which is the equation presented in Section 5.4.1 of Chapter 5. As in the case of full connectivity, in order to find the number of contacts that firm 1 receives from firm 2, and separate them from the number of contacts firm 1 receives from firm 3, one obtains:

$$I_{1,3} = \frac{\varepsilon - \varepsilon_{1,2}}{\varepsilon_{1,3} - \varepsilon_{1,2}} I_{1,j}$$

$$I_{1,2} = \frac{\varepsilon - \varepsilon_{1,3}}{\varepsilon_{1,2} - \varepsilon_{1,3}} I_{1,j}$$

Annex 2: Questionnaire for telecommunications business users

1. Company Profile

Name of the firm
Date of the interview
Name of the interviewed person
Position of the interviewed person
Address
Tel.
Principal products or service offered

Principal markets:

    Local ( ), regional ( ), national ( ), international ( )

	1988	1989	1990	1991
Turnover				
Net Profit				
Number of employees				
Number of skilled employees				
Number of employees in STAR related activities				
Location of headquarters in 1991				

2. Adopted Technologies

1. Which kinds of STAR infrastructures and services are you using and which are the most useful (Rank them in orde of importance 1 = most important)?

- infrastructure
  Optical Fibre Networks
  ISDN

Packet Switching Networks
LANs
Broadband Networks

- services
Videotex
Videoconference
Electronic mail
Electronic data interchange
Databanks
Telefax

2. Which of the following reasons was the most important in your decision to adopt the STAR technologies?

- high percentage of already networked firms
- importance of the network or the service for your business
- your suppliers or customers were connected
- no other networks or services connected
- low implementation costs
- low use costs
- image effect

3. How much do you use these services?

	Daily	Weekly	Monthly	Annual
Videotex				
Videoconference				
E-mail				
EDI				
Data banks				
Telefax				

4. For the services you have not adopted, which are the main reasons for this decision?

- high percentage of already networked firms
- importance of the network or the service for your business
- your suppliers or customers were connected
- no other networks or services connected
- low implementation costs
- low use costs
- image effect

5. Do you think of asking for other networks and services that you have not yet ued and, if so, which are the main reasons?

Networks:
  - a higher number of people connected
  - a higher number of suppliers and customers connected
  - a better geograhical distribution of the network
  - a reduction in the price of use
  - a reduction in the price of access
  - more recent availability of the entwork
  - technical progress in the network
  - increase in the efficiency of the network
  - good results from the previous adoption

Services:
  - a higher number of people connected
  - a higher number of suppliers and customers connected
  - a better geograhical distribution of the network
  - a reduction in the price of use
  - a reduction in the price of access
  - more recent availability of the network
  - technical progress in the network
  - increase in the efficiency of the network
  - good results from the previous adoption

6. Has the quality of communication improved since you adopted these technologies and if so in which sense?

    Yes   No

  - higher levels of reliability of communication technologies
  - higher volumes of communication
  - quicker communication
  - other (please specify)

7. Which are the technological innovations in the telecommunication technologies you would like to see implemented and why?

  - infrastructure
    Optical Fibre Networks
    ISDN
    Packet Switching Networks
    LANs

Broadband Networks

- services
  Videotex
  Videoconference
  Electroni mail
  Electronic data interchange
  Databanks

8. Are you satisfied with the quality of STAR technologies?

       Yes ( )   No ( )

If no, what are the major problems still existing in the use of these technologies

- too high costs for their implementation
- they are no longer free of charge services
- their use requires prfound organisational changes
- the number of subscribers using these services and networks is too low
- suppliers are not using these technologies
- competitors do not use these technologies
- other (please specify)

9. With the introduction of the new networks and services, has the intensity of your relationships

     Increased ( )?   Decreased ( )?   Remained Constant ( )?

10. If it has not increased, can you explain the reasons for this?

- None
- Cost of initial investment
- Lack of reliability on the technologies used
- Level of telecommunication tariff
- Lack of staff skills in their use
- Services are not relevant to business
- Your customers do not use them
- Your suppliers do not use them
- Too few subscribers in general
- Restricted geographical diffusion of networks and services

3. Advantages in the Adoption

11. What is the role played by the new telecommunication technologies in

                          Very important?   Important?   Modest?   None?
- your company turnover
- your company profit
- your total employment
- your staff and management employment
- your competitive position
  (national market share
   share of export on turnover)
- your innovative capacity
  (completely new products
   new versions of existing products)

Explain why

12. In particular, which are the major benefits that you obtained from the use of the STAR technologies?

    - increased efficiency of bureaucratic procedures
    - increased efficiency in the external relationships with:
      suppliers
      customers
    - cost saving
    - beating your competitors
    - presenting a better corporate image
    - job enrichment
    - greater quality of your final products/services
    - higher degree of innovativeness
    - other

13. What kind of innovation did you put in place which, according to you, has been facilitated in its implementation by the existence of STAR technologies?

    - product innovation (specify)
    - process innovation (specify)
    - managerial innovation (specify)
    - others

14. Did you change the number of your suppliers in the last five years, and if so, what are the main reasons?

 Increased ( )   Stable ( )   Decreased ( )

- change in your final products
- low quality of their products
- high costs of their products
- low quality of communications with them
- other (specify)

15. What was the role of the new telecommunication technologies in this process?

 Very important ( )   Important ( )   Modest ( )   None ( )

Explain why

16. Did you increase your geographical market in the last 5 years, and if yes where?

- no
- yes (specify where)

17. What was the role of the new telecommunication technologies in this process?

 Very important ( )   Important ( )   Modest ( )   None ( )

Explain why

18. Have you ever changed your organisational structure in the last five years in the following functions?

- finance
- sales
- production
- marketing
- distribution
- administration
- purchasing

- personnel
- R&D

19. In which functions did you face the most important changes?

- finance
- sales
- production
- marketing
- distribution
- administration
- purchasing
- personnel
- R&D

20. Were telecommunication technologies facilitating these processes?

      Yes No

21. If yes, how would you define the role of these technologies in these processes?

    Important ( ) Modest ( ) None ( )

4. Promotion Measures

22. How did you become aware of the existence of the STAR technologies?

- service demonstration centres
- entrepreneurial associations (unions)
- seminar and publicity campaign
- other firms in your region using them
- other

23. What role did the existence of service demonstration measures play in your decision to adopt telecommunication technologies?

    High ( ) Modest ( ) None ( )

Annex 3:
Definition of outliers

An analysis of the outliers on the basis of their mean and standard deviation shows quite clearly that two of the four outliers have a very high average performance index (case a in Table 1 below), much higher than the average performance index of the whole population (1533.3 against an average value of 253.91 in the case of the population). As Table 1 below shows, the value 1533.3 is even outside the minimum and maximum average productivity level of our population. Moreover, for the same two outliers the average value of the connectivity index is nothwithstanding much higher than the average level of the North sample (0.666), though in this case the value is contained between the minimum and the maximum average connectivity value of our population (excluding the outliers). We can easily argue that this firm is a large firm, or at least does not belong to the small-medium firms category on which the analysis is run. For the other two outliers (case b in Table 1 below), also in this case, in respect to the population, the average value of the connectivity index is much higher than the average level of the population (0.6 against 0.3), though it is contained in the range between the minimum and the maximum average values of the population (see Table 1 below). This analysis shows that these observations can be easily interpreted as outliers, and explains from a statistical point of view why the Pearson correlation coefficient improves once the analysis is run without these observations.

An analysis of the outliers is also run for the South of Italy. In this case, the average performance index is much higher than the average performance value of our population, excluding the outliers (3681.8 as opposed to 483.94). As Table 1 below shows, the average value of the connectivity index (0.333) is extremely near the average value of the population (0.1931), and for this reason it is difficult to treat these observations as outliers. It is not surprising, therefore, that the results obtained by running the correlation analysis without these observations do not lead to significant positive effects on the value of the Pearson correlation coefficient, as we will see hereafter.

Table 1
Values of outliers and population average values

			RESULTS FROM THE NORTH			
			Minimum	Maximum	Mean	Std Dev
Performance	Aver. Outliers Val.	a)	1066.7	2000	1533.3	659.9
Index	Aver. Outliers Val.		500	690.26	595.1	134.5
	Population Values		71.4	1066.7	253.9	195.3
Weighted	Aver. Outliers Val.	b)	0.666	0.666	0.666	0
Connectivity Index	Aver. Outliers Val.		0.167	0.167	0.167	0
	Population Values		0.167	0.9	0.334	0.198

		RESULTS FROM THE SOUTH			
		Minimum	Maximum	Mean	Std Dev
Performance	Outliers Values	3681.8	3681,8	3681,8	0.0
Index	Population Values	28.57	2000	483.9	603.5
Weighted	Outliers Values	0.33	0.33	0.33	0.0
Connectivity Index	Population Values	0	0.51	0.19	0.14

Annex 4:
Choice of cluster number

In the hierarchical clustering process, a sequence of cluster solutions is obtained with an "ideal" solution appearing for each possible number of clusters from n to 1. A second step of the cluster analysis is often to select an optimal number of clusters. To assist with the determination of the appropriate solution an optimality criterion is usually used. As the number of clusters g declines from n to 1 the cluster solution is evaluated by computing one or more available optimality criteria.

The simplest approach to cluster choice uses the value of the group proximity measure for the two groups joined at each step. As the process moves from step 1 to step $(n - 1)$, the value of the group proximity measure, say s, will increase (for dissimilarity measures).

In our specific case two approaches have been used for the selection of an appropriate value of g. The first measures the group proximity s with an R-squared (see the following table for both the South and the North of Italy). If a large change R-squared value occurs at some value of g then the solution (g + 1) immediately prior to this step should be chosen. In our specific case, for the South of Italy the first large change in the list of the R-squared values is when it changes from 0.78 (corresponding to 9 clusters) to 0.739 (corresponding to 8 clusters). Thus the choice of 9 clusters was made. For the North of Italy, the "jump" in the R-squared values occured between a value of 0.61 (6 clusters) and 0.55 (5 clusters). Thus, 6 clusters were chosen.

The second and alternative graphical approach used involves plotting the changes in s, as a function of the number of clusters (the so-called dendogram) (see graphs hereafter). When a drastic change occurs, the number of cluster associated with that point is indicative of the appropriate end to the clustering process.

Centroid hierarchical cluster analysis for the North of Italy

Eigenvalues of the Covariance Matrix

	Eigenvalue	Difference	Proportion	Cumulative
1	1.88025	0.714714	0.358000	0.35800
2	1.16554	0.160270	0.221918	0.57992
3	1.00527	0.457374	0.191403	0.77132
4	0.54789	0.109761	0.104319	0.87564
5	0.43813	0.223112	0.083420	0.95906
6	0.21502	.	0.040940	1.00000

Root-Mean-Square Total-Sample Standard Deviation = 0.935601
Root-Mean-Square Distance Between Observations = 3.241019

Number of Clusters	Clusters Joined		Frequency of New Cluster	Semipartial R-Squared	R-Squared	Approximate Expected R-squared	Cubic Clustering Criterion	Pseudo F	Pseudo t**2	Normalized Centroid Distance	Tie
34		37	49	2	0.000637	0.999363	.	.	47.56	.	0.147139	
33		55	62	2	0.000822	0.998541	.	.	42.78	.	0.167180	
32		38	51	2	0.000891	0.997650	.	.	41.09	.	0.174044	
31		45	63	2	0.000897	0.996753	.	.	40.93	.	0.174661	
30		41	47	2	0.001044	0.995709	.	.	40.01	.	0.188387	
29	CL31	CL33		4	0.002099	0.993610	.	.	33.32	2.44	0.188917	
28	CL29		48	5	0.002795	0.990814	.	.	27.97	2.20	0.243729	
27		44	56	2	0.001936	0.988878	.	.	27.36	.	0.256562	
26		46	66	2	0.002250	0.986629	.	.	26.56	.	0.276557	
25		42	52	2	0.002341	0.984288	.	.	26.10	.	0.282121	
24	CL27		67	3	0.003416	0.980872	.	.	24.53	1.76	0.295132	
23	CL28		50	6	0.005095	0.975777	.	.	21.97	3.08	0.322408	
22	CL24	CL23		9	0.012435	0.963342	.	.	16.27	5.10	0.325112	
21	CL30		53	3	0.004385	0.958956	.	.	16.35	4.20	0.334409	
20	CL21	CL26		5	0.008272	0.950684	.	.	15.22	3.23	0.342334	
19	CL20		59	6	0.005625	0.945059	.	.	15.29	1.41	0.338740	
18		43	61	2	0.004145	0.940914	.	.	15.92	.	0.375424	
17	CL25		68	3	0.005649	0.935265	.	.	16.25	2.41	0.379530	
16	CL19		60	7	0.008663	0.926602	.	.	15.99	2.01	0.414505	
15	CL16	CL22		16	0.048924	0.877678	.	.	10.25	11.47	0.459596	
14	CL18		69	3	0.009178	0.868499	.	.	10.67	2.21	0.483784	
13	CL17		64	4	0.010422	0.858078	.	.	11.08	2.61	0.486027	
12		39	CL15	17	0.013730	0.844348	.	.	11.34	1.90	0.498001	
11	CL12		54	18	0.014781	0.829566	.	.	11.68	1.93	0.515815	
10		36	65	2	0.008049	0.821518	.	.	12.79	.	0.523119	
9		40	58	2	0.011025	0.810493	.	.	13.90	.	0.612249	
8	CL11		57	19	0.023892	0.786601	.	.	14.22	2.96	0.654772	
7	CL34	CL14		5	0.033185	0.753416	0.708410	2.3677	14.26	7.13	0.685654	
6	CL7	CL8		24	0.134552	0.618864	0.668869	-2.1278	9.42	14.22	0.760175	
5	CL6	CL32		26	0.061146	0.557718	0.620628	-2.2710	9.46	4.27	0.750366	
4	CL10	CL5		28	0.075527	0.482191	0.558668	-2.3236	9.62	4.76	0.831480	
3	CL4	CL13		32	0.146036	0.336156	0.454558	-2.9856	8.10	8.64	0.842209	
2	CL3	CL9		34	0.165798	0.170358	0.303332	-3.0462	6.78	7.99	1.223668	
1	CL2		70	35	0.170358	0.000000	0.000000	0.0000	.	6.78	1.726633	

Centroid hierarchical cluster analysis
Distance between cluster centroids
North of Italy

Centroid hierarchical cluster analysis for the South of Italy

Eigenvalues of the Covariance Matrix

	Eigenvalue	Difference	Proportion	Cumulative
1	1.60095	0.487718	0.300371	0.30037
2	1.11323	0.198689	0.208865	0.50924
3	0.91454	0.031653	0.171586	0.68082
4	0.88289	0.261483	0.165648	0.84647
5	0.62140	0.424501	0.116588	0.96306
6	0.19690	.	0.036943	1.00000

Root-Mean-Square Total-Sample Standard Deviation = 0.942506
Root-Mean-Square Distance Between Observations = 3.264935

Number of Clusters	Clusters Joined		Frequency of New Cluster	Semipartial R-Squared	R-Squared	Approximate Expected R-squared	Cubic Clustering Criterion	Pseudo F	Pseudo t**2	Normalized Centroid Distance	Tie
32	15	21	2	0.000023	0.999977	.	.	1400.80	.	0.027146	
31	28	30	2	0.000051	0.999926	.	.	904.88	.	0.040256	
30	5	22	2	0.000098	0.999828	.	.	601.43	.	0.056087	
29	CL30	24	3	0.000727	0.999101	.	.	158.74	7.40	0.132106	
28	3	7	2	0.000786	0.998315	.	.	109.70	.	0.158608	
27	8	10	2	0.000803	0.997512	.	.	92.52	.	0.160289	
26	6	CL27	3	0.002002	0.995510	.	.	62.08	2.49	0.219202	
25	CL32	20	3	0.002614	0.992896	.	.	46.59	113.50	0.250458	
24	CL25	19	4	0.005262	0.987634	.	.	31.25	3.99	0.335047	
23	CL29	CL26	6	0.012379	0.975255	.	.	17.91	13.64	0.363371	
22	CL28	CL31	4	0.008532	0.966723	.	.	15.22	20.39	0.369483	
21	11	CL24	5	0.006926	0.959797	.	.	14.32	2.63	0.372176	
20	4	26	2	0.004469	0.955328	.	.	14.63	.	0.378169	
19	CL22	25	5	0.007346	0.947982	.	.	14.17	2.35	0.383310	
18	1	34	2	0.004628	0.943354	.	.	14.69	.	0.384839	
17	CL23	32	7	0.008163	0.935190	.	.	14.43	2.55	0.390365	
16	CL17	35	8	0.007373	0.927818	.	.	14.57	1.83	0.367170	
15	14	33	2	0.004951	0.922867	.	.	15.38	.	0.398035	
14	CL19	CL21	10	0.031086	0.891781	.	.	12.04	7.88	0.446037	
13	CL16	27	9	0.011360	0.880420	.	.	12.27	2.52	0.452203	
12	CL20	CL13	11	0.027642	0.852779	.	.	11.06	5.25	0.519879	
11	13	18	2	0.008471	0.844308	.	.	11.93	.	0.520648	
10	CL12	12	12	0.019460	0.824847	.	.	12.03	2.59	0.582815	
9	CL14	CL11	12	0.037270	0.787578	.	.	11.12	5.24	0.598155	
8	CL18	CL9	14	0.048213	0.739364	.	.	10.13	5.12	0.670813	
7	CL8	CL10	26	0.172017	0.567347	.	.	5.68	16.15	0.652647	
6	9	23	2	0.016922	0.550425	0.650672	-3.8309	6.61	.	0.735863	
5	CL7	31	27	0.034413	0.516012	0.595962	-2.7242	7.46	2.01	0.756170	
4	CL5	CL15	29	0.068328	0.447684	0.516285	-2.0580	7.84	3.95	0.766233	
3	CL4	CL6	31	0.129362	0.318322	0.415575	-2.5290	7.00	6.79	1.051792	
2	CL3	16	32	0.103260	0.215062	0.267110	-1.3525	8.49	4.54	1.305931	
1	CL2	29	33	0.215062	0.000000	0.000000	0.0000	.	8.49	1.883753	

Centroid hierarchical cluster analysis
Distance between cluster centroids
South of Italy

```
        1.9  1.8  1.7  1.6  1.5  1.4  1.3  1.2  1.1   1   0.9  0.8  0.7  0.6  0.5  0.4  0.3  0.2  0.1   0
S
K
D
D
   2    .....................................................................................................
  17    .....................................................................................................
   1    xxxxxxxxxxxxxxxxxxxxxxxxxxxxxxxxxxxxxxxxxxxxxxxxxxxxxxxxxxxxxxxxxxxxxxxxxxxxxxx.......................
  34    xxxxxxxxxxxxxxxxxxxxxxxxxxxxxxxxxxxxxxxxxxxxxxxxxxxxxxxxxxxxxxxxxxxxxxxxxxxxxxx.......................
   3    xxxxxxxxxxxxxxxxxxxxxxxxxxxxxxxxxxxxxxxxxxxxxxxxxxxxxxxxxxxxxxxxxxxxxxxxxxxxxxxxxxxxxxxxxx.............
   7    xxxxxxxxxxxxxxxxxxxxxxxxxxxxxxxxxxxxxxxxxxxxxxxxxxxxxxxxxxxxxxxxxxxxxxxxxxxxxxxxxxxxxx................
  28    xxxxxxxxxxxxxxxxxxxxxxxxxxxxxxxxxxxxxxxxxxxxxxxxxxxxxxxxxxxxxxxxxxxxxxxxxxxxxxxxxxxxxxxxxxxxxxxx......
  30    xxxxxxxxxxxxxxxxxxxxxxxxxxxxxxxxxxxxxxxxxxxxxxxxxxxxxxxxxxxxxxxxxxxxxxxxxxxxxxxxxxxxxxxxxxxxxx.......
  25    xxxxxxxxxxxxxxxxxxxxxxxxxxxxxxxxxxxxxxxxxxxxxxxxxxxxxxxxxxxxxxxxxxxxxxxxxxxxxxxxxx...................
  11    xxxxxxxxxxxxxxxxxxxxxxxxxxxxxxxxxxxxxxxxxxxxxxxxxxxxxxxxxxxxxxxxxxxxxxxxxxxxxxxxx....................
  15    xxxxxxxxxxxxxxxxxxxxxxxxxxxxxxxxxxxxxxxxxxxxxxxxxxxxxxxxxxxxxxxxxxxxxxxxxxxxxxxxxxxxxxxxxxxxxxxxxxx..
  21    xxxxxxxxxxxxxxxxxxxxxxxxxxxxxxxxxxxxxxxxxxxxxxxxxxxxxxxxxxxxxxxxxxxxxxxxxxxxxxxxxxxxxxxxxxxxxxxxx...
  20    xxxxxxxxxxxxxxxxxxxxxxxxxxxxxxxxxxxxxxxxxxxxxxxxxxxxxxxxxxxxxxxxxxxxxxxxxxxxxxxxxxx..................
  19    xxxxxxxxxxxxxxxxxxxxxxxxxxxxxxxxxxxxxxxxxxxxxxxxxxxxxxxxxxxxxxxxxxxxxxxxxxxxxxx......................
  13    xxxxxxxxxxxxxxxxxxxxxxxxxxxxxxxxxxxxxxxxxxxxxxxxxxxxxxxxxxxxxxxxxxxxxxxxx...........................
  18    xxxxxxxxxxxxxxxxxxxxxxxxxxxxxxxxxxxxxxxxxxxxxxxxxxxxxxxxxxxxxxxxxxxxxxx.............................
   4    xxxxxxxxxxxxxxxxxxxxxxxxxxxxxxxxxxxxxxxxxxxxxxxxxxxxxxxxxxxxxxxxxxxxxxxxxxxxxxxxxx..................
  26    xxxxxxxxxxxxxxxxxxxxxxxxxxxxxxxxxxxxxxxxxxxxxxxxxxxxxxxxxxxxxxxxxxxxxxxxxxxxxxxx....................
   5    xxxxxxxxxxxxxxxxxxxxxxxxxxxxxxxxxxxxxxxxxxxxxxxxxxxxxxxxxxxxxxxxxxxxxxxxxxxxxxxxxxxxxxxxxxxxxxxxx...
  22    xxxxxxxxxxxxxxxxxxxxxxxxxxxxxxxxxxxxxxxxxxxxxxxxxxxxxxxxxxxxxxxxxxxxxxxxxxxxxxxxxxxxxxxxxxxxxxx...
  24    xxxxxxxxxxxxxxxxxxxxxxxxxxxxxxxxxxxxxxxxxxxxxxxxxxxxxxxxxxxxxxxxxxxxxxxxxxxxxxxxxxxxx..............
   6    xxxxxxxxxxxxxxxxxxxxxxxxxxxxxxxxxxxxxxxxxxxxxxxxxxxxxxxxxxxxxxxxxxxxxxxxxxxxxxxxxxxx...............
   8    xxxxxxxxxxxxxxxxxxxxxxxxxxxxxxxxxxxxxxxxxxxxxxxxxxxxxxxxxxxxxxxxxxxxxxxxxxxxxxxxxxxx...............
  10    xxxxxxxxxxxxxxxxxxxxxxxxxxxxxxxxxxxxxxxxxxxxxxxxxxxxxxxxxxxxxxxxxxxxxxxxxxxxxxxxx..................
  32    xxxxxxxxxxxxxxxxxxxxxxxxxxxxxxxxxxxxxxxxxxxxxxxxxxxxxxxxxxxxxxxxxxxxxxxxxxxxxxxxxx................
  35    xxxxxxxxxxxxxxxxxxxxxxxxxxxxxxxxxxxxxxxxxxxxxxxxxxxxxxxxxxxxxxxxxxxxxxxxxxxx......................
  27    xxxxxxxxxxxxxxxxxxxxxxxxxxxxxxxxxxxxxxxxxxxxxxxxxxxxxxxxxxxxxxxxxxxxxxxx..........................
  12    xxxxxxxxxxxxxxxxxxxxxxxxxxxxxxxxxxxxxxxxxxxxxxxxxxxxxxxxxxxxxxxxxxxxxxx...........................
  31    xxxxxxxxxxxxxxxxxxxxxxxxxxxxxxxxxxxxxxxxxxxxxxxxxxxxxxxxxxxxxxxxxxxxxxx...........................
  14    xxxxxxxxxxxxxxxxxxxxxxxxxxxxxxxxxxxxxxxxxxxxxxxxxxxxxxxxxxxxxxxxxxxxxx............................
  33    xxxxxxxxxxxxxxxxxxxxxxxxxxxxxxxxxxxxxxxxxxxxxxxxxxxxxxxxxxxxxxxxxxx...............................
   9    xxxxxxxxxxxxxxxxxxxxxxxxxxxxxxxxxxxxxxxxxxxxxxxxxxxxxxxx...........................................
  23    xxxxxxxxxxxxxxxxxxxxxxxxxxxxxxxxxxxxxxxxxxxxxxxxxxxxxxx............................................
  16    xxxxxxxxxxxxxxxxxxxxxxxxxxxxxxxxxxxx...............................................................
  29    x..................................................................................................
```

References

Abernathy W. and Utterback J. (1978), "Patterns of Industrial Innovation", **Technology Review**, June, pp. 121-133

Adams R. and McCormick K. (1987), "Private Goods, Club Goods and Public Goods as a Continuum", **Review of Social Economy**, Vol. XLV, n. 2, October, pp. 192-199

Agresti A. (1990), **Categorical Data Analysis**, A Wiley Interscience Publication, John Wiley & Sons, New York

Allen D. (1988), "New Telecommunications Services: Network Externalities and Critical Mass", **Telecommunications Policy**, September, pp. 257-271

Allen D. (1990a), "Competition, Co-operation and Critical Mass in the Evolution of Networks", paper presented at the 8th Conference of the International Telecommunications Society, on "Telecommunications and the Challenge of Innovation and Global Competition", held in Venice, 18-21 March

Allen D. (1990b), "Break-up of the American Telephone Company: the Intellectual Conflict Before and Since", paper presented at the 8th International Conference of the International Telecommunications Society, on "Telecommunications and the Challenge of Innovation and Global Competition", held in Venice, 18-21 March

Allen D. (1992), "Telecommunication Policy Between Innovation and Standardisation", paper presented at the IXth International Conference of the International Telecommunication Society, held in Sophia Antipolis, 14-17 June

Amiel M. and Rochet J. (1987), "Concurrence entre Réseaux de Télécommunications: les Conséquences des Externalités Négatives", **Annales des Télécommunications**, 42, n. 11-12, pp. 642-649

Andersson A. and Johansson B. (1984), "Industrial Dynamics, Product Cycles and Employment Structures", working paper of International Institute for Applied Systems Analysis, Laxemburg, Austria

Andersson A. and Stroemqvist U. (1988), "The Emerging C-Society", in Batten D. and Thord R. (eds.), **Transportation for the Future**, Springer Verlag, Berlin, pp. 29-42

Andraisfai B. (1977), **Introductory Graph Theory**, Hilger, Bristol

Antonelli C. (1982), **New Information Technologies and Regional Economic Development**, Report to the Commission of the European Communities, Directorate General for Regional Affairs, Brussels

Antonelli C. (ed.) (1988), **New Information Technology and Industrial Change: the Italian Case**, Kluwer Academic Publisher Books, New York

Antonelli C. (1989), "The Diffusion of Information Technology and the Demand for Telecommunication Services", **Telecommunications Policy**, September, pp. 255-264

Antonelli C. (1990), "Induced Adoption and Externalities in the Regional Diffusion of Information Technology", **Regional Studies**, vol. 24, n.1, pp. 31-40

Antonelli C. (1991), The International Diffusion of Advanced Telecommunications: Opportunities for Developing Countries, OECD Publication

Antonelli C. (ed.) (1992), **The Economics of Information Networks**, Elsevier Publisher, Amsterdam

Antonelli C. (1993), "Reti: Varietà e Complementarietà", paper presented at the seminar on **Comunicare nella Metropoli** organised by Tecnopolis and Sip, held in Bari, 3-4 June 1993

Antonelli C. and Gottardi G. (1988), "Interazioni tra Produttori e Utilizzatori nei Processi di Diffusione Tecnologica", **L'Industria**, n.4, pp. 599-636

Antonelli C, Petit P. and Tahar G. (1989), "Technological Diffusion and Investment Behaviour: the Case of the Textile Industry", **Weltwirtschaftlisches Archiv**, n. 125, pp. 782-803

Arcangeli F. (1984), "Verso un Paradigma Economico e Spaziale per l'Analisi della Diffusione delle Innovazioni nella Produzione e nel Consumo", in Camagni R., Cappellin R. and Garofoli G. (eds.) **Cambiamento Tecnologico e Diffusione Territoriale**, Franco Angeli, Milan, pp. 87-108

Arcangeli F, Dosi G., Moggi M. (1987), "Patterns Diffusion of Electronics Technologies", paper presented at the GERTTD-AMES Conference on Automatisation Programmable et Conditions d'Usage du Travail Paris, April

Aydalot Ph. (ed.) (1986), **Milieux Innovateurs en Europe**, GREMI, Paris

Aydalot Ph. and Keeble D. (eds.) (1988), **High Technology Industry and Innovative Environments**, Routledge, London

Ayres R.U. and Miller S.M. (1983), **Robotics: Applications and Social Implications**, Ballinger Publishing Company, Cambridge Massachusetts

Bagnasco A. (1977), **Le Tre Italie**, Il Mulino, Bologna

Bakis H., Abler R. and Roche M. (eds.) (1993), **Corporate Networks, International Telecommunicatios and Interdependence**, Belhaven Press, London

Bar F., Borrus M. and Coriat B. (1989), "Information Networks and Competitive Advantages: the Issues for Government Policy and Corporate Strategy", OECD-BRIE Telecommunications User Project, Paris

Becattini G. (ed.) (1987), **Mercato e Forze Locali: il Distretto Industriale**, Il Mulino, Bologna

Bell D. (1973), **The Coming of Post-industrial Society**, Basic Books, New York

Behzad M. and Chartrand G. (1971), **Introduction to the Theory of Graphs**, Allyn and Bacon, Boston

Belussi F. (1990), "Innovation Diffusion, Innovation Acquisition and Path Dependency in Traditional Sectors: an Empirical Investigation for the Veneto Region of Italy", in Ciciotti E., Alderman N. and Thwaites A. (eds.), **Technological Change in a Spatial Context**, Springer Verlag, Berlin, pp. 285-316

Ben-Akiva M. and Lerman S.R. (1985) **Discrete Choice Analysis: Theory and Application to Travel Demand,** MIT Press, Cambridge, Massachusetts

Bental B. and Spiegel M. (1990), "Consumption Externalities in Telecommunication Services", in de Fontenay M. and Sibley D. (eds.), **Telecommunications Demand Modelling**, Elsevier Science Publisher, pp. 415-432

Berglas E. (1976), "On the Theory of Clubs", **The American Economic Review**, May, vol. 66, n. 2, pp. 116-121

Berglas E. (1980), "The Market Provision of Club Goods Once Again", **Journal of Public Economics**, vol. 15, pp. 389-393

Biehl D. (1986), **The Contribution of Infrastructure to Regional Development,** Regional Policy Division, European Community, Brussels

Bishop Y.M.M., Fienberg S. E. and Holland P. W. (1977), **Discrete Multivariate Analysis: Theory and Practice**, MIT Press, Massachusetts

Blaas E. (1991), Evaluation of Alternative Methodological Approaches for Assessing the Regional Macro-Economic Effects of the Community's Regional Action, Research Report, Free University of Amsterdam

Blankart C. and Knieps G. (1992), "The Critical Mass Problem in a Dynamic World: Theory and Applications to Telecommunication", in Klaver F. and Slaa P. (eds.), **Telecommunications: New Signposts to Old Roads**, IOS Press, Amsterdam, pp. 55-63

Blommestein H. and Nijkamp P. (1983), "Causality Analysis in Soft Spatial Econometric Models", **Papers of the Regional Science Association**, Vol. 51, pp. 65-77

Blum U. (1982), "Effects of Transportation Investments on Regional Growth", **Papers of the Regional Science Association**, vol. 49, pp. 151-168

Boadway R. (1980), "A Note on the Market Provision of Club Goods", **Journal of Public Economics**, vol. 13, pp. 131-137

Boeckhout I. and Molle W. (1982), "Technological Change, Location Patterns and Regional Development", Netherland Economic Institute Report PRESTO-Project, Rotterdam

Boitani A. and Ciciotti E. (1990), "Patents as Indicators of Innovative Performances at the Regional Level", in Cappellin R. and Nijkamp P. (eds.), **The Spatial Context of Technological Development**, Avebury, Aldershot, pp. 139-166

Bollen K.A. (1989), **Structural Equations with Latent Variables**, John Wiley and Sons., New York

Boudeville J.R. (1968), **L'Espace et le Pole de Croissance**, Presses Universitaires de France, Paris

Boyce D., Nijkamp P. and Shefer D. (eds.) (1991), **Regional Science: Retrospect and Prospect**, Springer Verlag, Berlin

Bradley S. and Hausman J. (eds.) (1989), **Future Competition in Telecommunications**, Harvard Business School Press

Brock G. (1981), **The Telecommunications Industry: the Dynamic of Market Structure**, Cambridge, Harvard University Press

Brosio G. (1986), **Economia e Finanza Pubblica**, La Nuova Italia Scientifica, Rome

Brotchie J., Batty M., Hall P. and Newton P. (1991) (eds.), **Cities of the 21st Century**, Halstead Press, Longman, Cheshire

Brown L. (1981), **Innovation Diffusion: A New Perspective**, Methuen, London

Bruinsma F., Nijkamp P. and Rietveld P. (1989), "Employment Impacts of Infrastructure Investments: a Case Study for the Netherlands", in Peschel K. (ed.), **Infrastructure and the Space Economy**, Springer Verlag, Berlin, pp. 209-226

Brusco S. (1982), "The Emilian Model: Productive Decentralisation and Social Integration", **The Cambridge Journal of Economics**, n. 6, pp. 167-184

Buchanan J. (1965), "An Economic Theory of Clubs", **Economica**, February, pp.1-14

Burns R. and Stalker G. (1961), **The Management of Innovation**, Tavistock Publications, London

Bushwell R. and Lewis E. (1970), "The Geographic Distribution of Industrial Research Activity in the U.K.", **Regional Studies**, vol. 4, pp. 297-306

Button K. and Gillingwater D. (eds.) (1986a), **Transport, Location and Spatial Policy**, Gower, Aldershot

Button K. and Gillingwater D. (1986b), **Future Transport Policy**, Croom Helm, London

Cabral L. and Leite A. (1992), "Network Consumption Externalities: the Case of the Portuguese Telex Service", in Antonelli C. (ed.), **The Economics of Information Networks**, Elservier Publisher, pp. 129-140

Cainarca G., Colombo M. and Mariotti S. (1989), "An Evolutionary Pattern of Innovation Diffusion: the Case of Flexible Automation", **Research Policy**, vol. 18, pp. 59-86

Camagni R. (ed.) (1984), **Il Robot Italiano**, Il Sole 24Ore, Milan

Camagni R. (1985), "Spatial Diffusion of Pervasive Process Innovation", **Papers of the Regional Science Association**, vol. 58, pp. 83-95

Camagni R. (1986), "La Diffusione di Processi di Automazione Flessibile: il Caso dell'Industria Lombarda", **Economia e Politica Industriale**, n. 52, pp. 153-196

Camagni R. (1988), "Functional Integration and Locational Shifts in The New Technology Industry", in Aydalot Ph. and Keeble D. (eds.) **High Technology Industry and Innovative Environments**, Routledge, London, pp. 48-64

Camagni R. (ed.) (1991a), **Computer Network: Mercati e Prospettive delle Tecnologie di Comunicazione**, Etas Libri, Milano

Camagni R. (ed.) (1991b), **Innovation Networks: Spatial Perspectives**, Belhaven-Pinter, London

Camagni R. (1991c), "Regional Deindustrialisation and Revitalisation Processes in Italy", in Rodwin L. and Sazanami H. (eds.), **Industrial Change and Regional Economic Transformation**, Harper Collins, pp.137- 167

Camagni R. (1991d), "Interregional Disparities in the European Community: Structure and Performance of Objective 1 Regions in the 1980's", paper presented at the North American Regional Science Conference, New Orleans, 6-9 November

Camagni R. (1992a), "The Spatial Implications of Technological Diffusion and Economic Restructuring in Europe: the Italian Case", in **Ekistics**, Special Issue on Urban Networking in Europe, vol. 58, September-December, pp. 330-335

Camagni R. (1992b), **Economia Urbana**, La Nuova Italia Scientifica, Rome

Camagni R. (1993), "La Valorizzazione dei Fattori Locali del Processo Innovativo: Università, Imprese, Governo Locale e Centrale", in APSTI (Associazione Parchi Scientifici e Tecnologici Italiani), **Parchi e Poli Scientifici e Tecnologici**, CUEN Publisher, Naples, pp. 33-42

Camagni R. (1994), **La Teoria dello Sviluppo Regionale: un Approccio Europeo**, La Nuova Italia Scientifica, Rome

Camagni R. and Capello R. (1989), "Scenari di Sviluppo della Domanda di Sistemi di Telecomunicazione in Italia", **Finanza, Marketing e Produzione**, n.1, March, pp. 87-138

Camagni R. and Capello R. (1990), "Tecnologia e Organizzazione in Banca", **Sviluppo e Organizzazione**, n. 122, November/December, pp. 109-125

Camagni R. and Capello R. (1991), "Le Caratteristiche delle Nuove Tecnologie di Comunicazione e loro Interazione con la Domanda", in Camagni R. (ed.) **Computer Network: Mercati e Prospettive delle Tecnologie di Telecomunicazione**, Etas Libri, Milan, pp. 3-42

Camagni R. and Capello R. (1993), "Nuove Tecnologie di Comunicazione e Cambiamenti nella Localizzazione delle Attività Industriali", in Lombardo S. (ed.) **Nuove Tecnologie dell'Informazione e Sistemi Urbani**, La Nuova Italia Scientifica, Rome, forthcoming

Camagni R., Capello R. and Scarpinato M. (1993), "Scenarios for the Italian Telecommunication Market: Private and Public Strategies", in **Telecommunications Policy**, vol. 17, pp. 27-48

Camagni R. and Cappellin R. (1984), "Cambiamento Strutturale e Dinamica della Produttività nelle Regioni Europee", in Camagni R., Cappellin R. and Garofoli G. (eds.), **Cambiamento Tecnologico e Diffusione Territoriale**, pp.131-337

Camagni R., Cappellin R. and Garofoli G. (eds.) (1984), **Cambiamento Tecnologico e Diffusione Territoriale**, Franco Angeli Editore, Milan

Camagni R. and De Blasio G. (ed.) (1993), **Le Reti di Città**, Franco Angeli Editore, Milan

Camagni R. and Pattarozzi M. (1984), "L'evoluzione Territoriale di un'Innovazione di Processo e di Prodotto: il Caso della Robotica in Italia", **L'Industria**, n.4, October-December, pp. 491-525

Camagni R. and Rabellotti R. (1988), "Innovation and Territory: the Milan Information Technology Field", in Giaoutzi M. and Nijkamp P. (eds.) (1988), **Informatics and Regional Development**, Avebury, Aldershot, pp.215-234

Camagni R. and Rabellotti R. (1990), "Advanced Technology Policies and Strategies in Developing Regions", in Ewers H. and Allesch J. (eds.) **Innovation and Regional Development**, De Gruyter, Berlin, pp. 235-264

Camagni R. and Rabellotti R. (1992), "Technology and Organisation in the Italian Textile-clothing Industry", **Entrepreneurship and Regional Development**, n. 4, pp. 271-285

Capello R. (1988), "La Diffusione Spaziale dell'Innovazione: il Caso del Servizio Telefonico", **Economia e Politica Industriale**, n. 58, pp. 141-175

Capello R. (1991a), "Dinamica Tecnologica e Dinamica Istituzionale: Verso Nuovi Comportamenti d'impresa nel Settore delle Telecomunicazioni", **Economia e Politica Industriale**, n. 68, pp. 261-277

Capello R. (1991b), "L'assetto Istituzionale nel Settore delle Telecomunicazioni", in Camagni R. (ed.), **Computer Networks: Mercati e Prospet-tive delle Tecnologie di Telecomunicazione**, Etas Libri, Milan, pp.163-260

Capello R. (1993), "Verso Nuovi Sistemi Spaziali e Industriali: il Ruolo delle Nuove Tecnologie", in Lombardo S. and Preto G. (eds.), **Innovazione e Trasformazioni della Città**, Franco Angeli Editore, Milan, pp. 297-326

Capello R., Taylor J. and Williams H. (1990), "Computer Networks and Competitive Advantage in Building Society", **International Journal of Information Management**, vol. 10, pp. 54-66

Capello R. and Williams H. (1990), "Nuove Strategie d'Impresa, Nuovi Sistemi Spaziali e Nuove Tecnologie dell'Informazione come Strumenti di Riduzione dell'Incertezza", **Economia e Politica Industriale**, n. 67, pp. 43-72

Capello R. and Williams H. (1992), "Computer Network Trajectories and Organisational Dynamics: a Cross-national Review", in Antonelli C. (ed.), **The Economics of Information Networks**, North Holland, Amsterdam, pp. 347-362

Cappellin R. (1983), "Osservazioni sulla Distribuzione Inter and Intraregionale delle Attività Produttive", in Fuà G. and Zacchia C. (eds.) **Industrializzazione senza Fratture**, Il Mulino, Bologna, pp. 241-272

Cappellin R. and Nijkamp P. (eds.) (1990), **The Spatial Context of Technological Development**, Avebury, Aldershot

Charles D., Monk P. and Sciberras E. (1989), **Technology and Competition in the International Telecommunications Industry**, Pinter Publisher, London

Christofides N. (1975), **Graph Theory: An Algorithmic Approach**, Academic Press, New York

Church J. and Gandal N. (1992), "Network Effects, Software Provision, and Standardisation", **The Journal of Industrial Economics**, Vol. XL, no. 1, March, pp. 85-123

Ciborra C. (1989), **Le Tecnologie di Coordinamento**, Franco Angeli Editore

Ciciotti E. (1984), "Innovazione e Sviluppo Regionale: Alcune Considerazioni sulle Implicazioni di Politica Economica", in Camagni R., Cappellin R. and Garofoli G. (eds.), **Cambiamento Tecnologico e Diffusione Territoriale**, Franco Angeli Editore, Milan, pp. 295-316

Ciciotti E., Alderman N. and Thwaites A. (eds.) (1990), **Technological Change in a Spatial Context**, Springer Verlag, Berlin

Clark C. (1940), **The Conditions of Economic Progress**, London, Macmillan

Clark N. (1972), "Science Technology and Regional Economic Development", **Research Policy**, n. 1, pp. 296-319

Coase R. H. (1937), "The Nature of the Firm", **Economica**, November, pp. 386-403

Coase R. H. (1960), "The Problem of Social Cost", **The Journal of Laws and Economics**, vol. 3, pp. 1-44

Colombino U. (1993), "Stima di un Sistema di Domanda di Servizi Telefonici con Uso di Dati Individuali", research project sponsored by Sip and developed at the Dipartimento di Economia e Produzione, Politecnico di Milano, Piazza Leonardo da Vinci 32, Milan

Colombo M. and Mariotti S. (1985), "Note Economiche sull'Automazione Flessibile", **Economia e Politica Industriale**, n. 48, pp. 61-96

Coriat B. (1981), "Robots et Automates dans l'Industrie de Serie. Esquisse d'une "Economie" de la Robotique d'Atelier", in Adelfi A. (ed.) **Les Mutations Technologiques**, Economica, Paris

Cornes R. and Sandler T. (1986), **The Theory of Externalities, Public Goods and Club Goods**, Cambridge University Press, Cambridge

Cowan D. and Waverman L. (1971), "The Interdependence of Communications and Data Processing Issues in Economics of Integration and Public Policy", **Bell Journal of Economics and Management Science**, 2, n. 2, Autumn, pp. 657-677

Cowen T. (ed.) (1988), **The Theory of Market Failure: a Critical Examination**, George Mason University Press

Crandall R. and Flamm K. (eds.) (1989), **Changing the Rules: Technological Change, International Competition and Regulation in Communications**, The Brooking Institution

Curien N. and Gensollen M. (1987), "A Functional Analysis of the Network: a Prerequisite for Deregulating the Telecommunications Industry", **Annales des Télécommunications**, 42, n. 11-12, pp. 629-641

Curien N. and Gensollen M. (1992), **Economie des Télécommunications**, ENSPTT, Economica, Paris

Davelaar E. J. (1991), **Regional Economic Analysis of Innovation and Incubation**, Avebury, Aldershot

Davelaar E. J. and Nijkamp P. (1990), "Industrial Innovation and Spatial Systems: the Impact of Producer Services", in Ewers H. and Allesch J. (1990) (eds.) **Innovation and Regional Development**, de Gruyter, Berlin, pp.83-122

Davelaar E. J. and Nijkamp P. (1992), "Operational Models on Innovation and Spatial Development: a Case-Study for the Netherlands", **Journal of Scientific and Industrial Research**, vol. 51, march, pp. 273-284

David P. (1985), "Clio and the Economics of Qwerty", **AEA Papers and Proceedings**, vol. 75, n. 2, pp. 332-337

David P. (1992), "Information Network Economics: Externalities, Innovation and Evolution", in Antonelli C. (ed.), **The Economics of Information Networks**, North Holland, Amsterdam, pp. 103-106

David P. (1993), "Compatibility Standards, Competition and the Development of Telecommunications Networks", paper presented at the seminar on

Comunicare nella Metropoli organised by Tecnopolis and Sip, held in Bari, 3-4 June 1993

Davies D. (1979), **The Diffusion of Process Innovations**, Cambridge University Press, Cambridge

Decoster E. and Tabariés M. (1986), "L'Innovation dans un Pole Scientifique: le Cas de la Cité Scientifique Ile de France Sud", in Aydalot Ph. (ed.) (1986), **Milieux Innovateurs en Europe**, GREMI, Paris, pp.79-100

Dhebar A. and Oren S. (1986), "Dynamic nonlinear Pricing in Networks with Interdependent Demand", **Operation Research**, vol. 34, n. 3, May-June, pp. 384-394

Domencich T. A. and Mc Fadden D. (1975), **Urban Travel Demand: a Behavioural Analysis**, North Holland, Amsterdam

Dosi G. (1982), "Technological Paradigms and Technological Trajectories: a Suggested Interpretation of the Determinants and Directions of Technical Change", **Research Policy**, vol. 11, pp. 147-162

Dosi G. Freeman C., Nelson R., Silverberg G. and Soete L. (eds.) (1988), **Technical Change and Economic Theory**, Pinter Publisher, London

Economides N. (1989), "Desirability of Compatibility in the Absence of Network Externalities", **American Economic Review**, vol. 78, n. 1, pp. 108-121

Economides N. (1992), "Network Externalities and Innovation to Enter, paper presented at the IXth International Conference of the International Telecommunication Society, held in Sophia Antipolis, 14-17 June

Ewers H. and Allesch J. (eds.) (1990), **Innovation and Regional Development**, de Gruyter, Berlin

Ewers H. and Wettman R. (1980), "Innovation Oriented Regional Policy", **Regional Studies**, vol. 14, pp. 161-179

Ewers H., Wettman R., Bade F., Kleine J. and Kirst H. (1979), **Innovationsorientierte Regionalpolitik**, Research Report, Wissenshaftszentrum, Berlin

Farrell J. and Saloner G. (1985), "Standardisation, Compatibility, and Innovation", **Rand Journal of Economics**, vol. 16, n. 1, Spring, pp. 70-83

Farrell J. and Saloner G. (1986), "Installed Base and Compatibility: Innovation, Product Preannouncements and Predation", **The American Economic Review**, vol. 76, n. 5, pp. 940-955

Fischer M. and Nijkamp P. (1985), "Developments in Explanatory Discrete Spatial Data and Choice Analysis", **Progress in Human Geography**, n. 9, pp. 515-551

Fischer M., Maggi R. and Rammer C. (1992), "Stated Preference Models of Contact Decision Behaviour in Academia", **Papers in Regional Science: the Journal of RSAI**, n. 71, vol. 4, pp. 359-371

Foreman-Peck J. and Mueller J. (1988), **The Spectrum of Alternatives of Market Configurations in European Telecommunications**, research study Berlin-Newcastle, April

Fornengo G. (1988), "Manufacturing Networks: Telematics in the Automobile Industry", in Antonelli (ed.), **New Information Technology and Industrial Change: the Italian Case**, Kluwer Academic Publisher Books, New York, pp. 33-56

Freeman C. (1987), **Technology Policy and Economic Performance**, Frances Pinter, London

Freeman C., Clark J. and Soete L. (1982), **Unemployment and Technical Innovation**, Frances Pinter, London

Fuà G. and Zacchia C. (eds.) (1983), **Industrializzazione senza Fratture**, Il Mulino, Bologna

Fuller W.A. (1987), **Measurement Errors Models**, John Wiley and Sons, New York

Gambardella A. (1985), "Innovazioni Tecnologiche e Comportamenti Soggettivi nell'Odierna Evoluzione Telematica", **Economia e Politica Industriale**, n. 7, pp. 59-94

Garnham N. (ed.) (1989) **European Telecommunications Policy Research**, Springfield Publisher, Amsterdam

Garofoli G. (1981), "Lo Sviluppo delle Aree Periferiche nell'Economia Italiana degli Anni Settanta", **L'Industria**, n. 3, pp. 391-404

Gent H. van and Nijkamp P. (1988), "Mobility, Transportation and Development", **I.A.T.T.S. Research**, vol. 11, n.1, pp. 62-68

Giaoutzi M. and Nijkamp P. (eds.) (1988), **Informatics and Regional Development**, Avebury, Aldershot

Gibbs D. and Thwaites A. (1985), "The Location and Potential Mobility of Research and Development Activity: A Regional Perspective", paper presented at the 25th European Conference of the Regional Science Association

Gillespie A. (1991), "Advanced Communications Networks, Territorial Integration and Local Development", in Camagni R. (ed.), **Innovation Networks**, Belhaven Press, London, pp. 214-229

Gillespie A. and Hepworth M. (1986), "Telecommunications and Regional Development in the Information Economy" **Newcastle Studies of the Information Economy**, CURDS, Newcastle University, n. 1, October

Gillespie A. and Williams H. (1988), "Telecommunications and the Reconstruction of Regional Comparative Advantage", **Environment and Planning A**, vol.20 pp. 1311-1321

Gillespie A., Goddard J, Robinson F., Smith I. and Thwaites A. (1984), **The Effects of New Technology on the Less-favoured Regions of the Community**, Studies Collection, Regional Policy Series, EEC, n. 23

Gillespie A., Goddard J., Hepworth M. and Williams H. (1989), "Information and Communications Technology and Regional Development: an Information Economy Perspective", **Science, Technology and Industry Review**, no. 5, April, pp. 86-111

Goddard J (1980), "Technology Forecast in a Spatial Context", **Futures**, April, pp 90-105

Goddard J. (1985), "Effetti delle Nuove Tecnologie dell'Informazione sulla Struttura Urbana", IReR Progetto Milano, Tecnologie e Sviluppo Urbano, Franco Angeli Editore, pp. 71-114

Goddard J., Charles D., Howells J. and Thwaites A. (1987), Research and Technological Development in the Less Favoured Regions of the Community - STRIDE, Commission of the European Communities, Document, Luxemburg

Goddard J. and Thwaites A. (1980), "Technological Change and the Inner City", The Inner City in Context, Social Science Research Council, Newcastle

Griffin J. (1982), "The Welfare Implications of Externalities and Price Elasticities for Telecommunications Pricing", **The Review of Economics and Statistics**, pp. 59-66

Griguolo S. and Reggiani A. (1985), "Modelli di Scelta tra Alternative Discrete: Alcune Note Introduttive", **Archivio di Studi Urbani e Regionali**, vol. 22, pp. 47-86

Griliches Z. (1957), "Hybrid Corn: an Exploration in the Economics of Technological Change", **Econometrica**, vol. 25, n. 4, October, pp. 501-525

Griliches Z. and Intriligator D. (eds.) (1983), **Handbook of Econometrics**, North Holland, Amsterdam

Grossman G. and Helpman E. (1991), **Innovation and Growth in the Global Economy**, The MIT Press, Cambridge, Massachusetts

Hägerstrand T. (1967), "Aspects of the Spatial Structure of Social Communication and the Diffusion of Innovation", **Papers of the Regional Science Association**, n. 16, pp. 27-42

Hayashi K. (1988), "The Economics of Networking, Implications for Telecommunications Liberalisation", paper presented at the 7th International Telecommunications Conference of the International Telecommunication Society, held in Boston

Hayashi K. (1989), "IT-based Networks, The Phenomenon of Networking and its Likely Innovation Dynamics", paper presented at OECD-ICCP Forum, October

Hayashi K. (1992), "From Network Externalities to Interconnection: the Changing Nature of Networks and Economy", in Antonelli C. (ed.), **The Economcs of Information Networks**, North-Holland, Amsterdam, pp. 195-216

Hepworth M. (1987), "Information Technology as Spatial Systems", in **Progress in Human Geography**, 11, pp. 157-180

Hepworth M. (1989), **The Geography of the Information Economy**, Belhaven Press, London

Hepworth M. and Waterson M. (1988), "Information Technology and the Spatial Dynamics of Capital", **Information Economics and Policy**, vol. 3, pp. 148-163

Hirsch S. (1967), **Location of Industry and International Competitiveness**, Clarendon Press, London

Howells J. (1984), "The Location of Research and Development: Some Observations and Evidence from Britain", **Regional Studies**, vol. 18, pp. 13-29

Howells J. (1990), "The Location and Organisation of Research and Development: New Horizons", **Research Policy**, 19, pp. 133-146

Jelinek M. and Golhar J.D. (1983), "The Interface between Strategy and Manufacturing Technology", **Columbia Journal of World Business**, Spring

Jonscher C. (1983), "Information Resources and Economic Productivity", **Information Economics and Policy**, vol. 1, pp. 13-35

Kamann D.-J. (1986), "Industrial Organisation, Innovation and Employment", in Nijkamp P. (ed.), **Technological Change, Employment and Spatial Dynamics,** Springer-Verlag, Berlin, pp. 131-154

Kamann D-J. and Nijkamp P. (1991), "Technogenesis", in **Technological Forecasting and Social Change**, Special Issue, vol. 39, n. 1-2, March/April, pp.45-66

Karshenas M. and Stoneman P. (1990), "Rank, Stock, Order and Epidemic Effects in the Diffusion of New Process Technologies: An Empirical Model", Working Paper n. 358, Warwick University, Conventry CV4 7AL, Great Britain

Katz M. and Shapiro C. (1985), "Network Externalities, Competition and Compatibility", **The American Economic Review**, vol. 75, n. 3, pp. 424-440

Katz M. and Shapiro C. (1986), "Technology Adoption in the Presence of Network Externalities", **Journal of Political Economy**, pp. 822-841

Katz M and Shapiro C. (1992), "Product Introduction with Network Externalities", **The Journal of Industrial Economics**, vol. XL, n. 1, pp.55- 83

Kellerman A. (1993), **Telecommunications and Geography**, Belhaven Press, London

Klaver F. and Slaa P. (1992), **Telecommunication: New Signpost to Old Roads**, IOS Press, Amsterdam

Leonardi G. (1985), "Equivalenza Asintotica fra la Teoria delle Utilità Casuali e la Massimizzazione dell'Entropia", in Reggiani A. (ed.), **Territorio e**

Trasporti: Modelli Matematici per l'Analisi e la Pianificazione, Franco Angeli, Milano, pp. 29-66

Liebowitz S. and Margolis S. (1990), "The Fable of the Keys", **Journal of Law and Economics**, vol. 33, pp. 1-25

Linhart P., Radner R. and Tewari R. (1992), "On the Market of Data Networking Products", in Antonelli C. (ed.), **The Economics of Information Networks**, North Holland, Amsterdam, pp. 141-156

Littlechild S. (1975), "Two-part Tariffs and Consumption Externalities", **The Bell Journal of Economics**, vol. 6, n.2, pp. 661-670

Loehlin J. C. (1987), **Latent Variable Models**, Hillsdale, New York

Machlup F. (1962), **The Production and Distribution of Knowledge in the United States**, Princeton University Press, New York

Maggi R., and Nijkamp P. (1992), "Missing Networks in Europe", **Transport Review**, vol. 12, n. 4, pp. 311-321

Maillat D. and Perrin J-C. (eds.) (1992), **Entreprises Innovatrices et Réseaux Locaux**, EDES, Neuchatel

Maillat D., Quévit M. and Senn L. (eds.) (1993), **Réseaux d'Innovation et Milieux Innovateurs: un Pari pour le Developpement Régional**, EDES, Neuchatel

Malecki E. (1979), "Locational Trends in R&D by Large U.S. Corporations 1965-1977", **Economic Geography**, vol. 55, pp. 309-323

Malecki E. (1980), "Corporate Organisation of R&D and the Location of Technological Activities", **Regional Studies**, vol. 55, pp. 309-323

Malecki E. and Varaija P. (1986), "Innovation and Changes in Regional Structure", in Nijkamp P. (ed.), **Technological Change, Employment and Spatial Dynamics,** Springer-Verlag, Berlin, pp. 625-645

Mansell R. (1990), "Rethinking the Telecommunication Infrastructure: the New 'Black Box'", **Research Policy**, vol. 19, pp. 501-515

Mansfield E. (1961), "Technological Change and the Rate of Imitation", **Econometrica**, vol. 29, n. 4, pp. 741-766

Mansfield E. (1968), **The Economics of Technological Change**, Norton, New York

Markus M. (1987), "Towards a Critical Mass Theory of Interarctive Media: Universal Access, Interdependence and Diffusion", **Communication Research**, vol. 14, pp. 491-511

Markus M. (1992), "Critical Mass Contingencies for Telecommunication Consumers", in Antonelli C. (ed.), **The Economics of Information Networks**, North Holland, Amsterdam, pp. 431-450

Marshall A. (1919), **Industry and Trade**, Macmillan, London

Marshall C. (1971), **Applied Graph Theory**, John Wiley, New York

Matutes C. and Regibeau P. (1988), "Mix and Match: Product Compatibility without Network Externalities", **Rand Journal of Economics**, vol. 19, n. 2, Summer, pp. 221-234

McFadden D. (1983), "Econometric Analysis of Qualitative Response Models", in Griliches Z. and Intriligator D. (eds.) (1983), **Handbook of Econometrics**, North Holland, Amsterdam, pp. 1396-1450

Meade J.E. (1954) "External Economies and Diseconomies in a Competitive Situation", **Economic Journal**, vol. 62, pp. 143-151

Metcalfe J. (1981), "Impulse and Diffusion in the Study of Technological Change", **Futures**, vol. 13, n. 5, (special issue), pp. 347-359

Mignolet M. (1983), **Les Processus de Diffusion des Innovations dans l'Espace et le Redéploiment Economique Régional**, Press Universitaire de Namur, Geneva

Mishan E.J. (1971), "The Postwar Literature on Externalities: an Interpretative Essay", **The Journal of Economic Literature**, vol. 9, March, pp. 1-28

Modigliani F. (1958), "New Developments on the Oligopoly Front", **Journal of Political Economy**, vol. 66, n. 3, pp. 215-233

Molle W. (1984), "Potenziali Regionali di Innovazione nella Comunità Europea", in Camagni R., Cappellin R. and Garofoli G. (eds.) (1984), **Cambiamento Tecnologico e Diffusione Territoriale**, Franco Angeli Editore, pp. 109-130

Momigliano F. (1984), "Revisione di Modelli Interpretativi delle Determinanti ed degli Effetti dell'Attività Innovativa, dell'Aggregazione Spaziale dei Centri di R&S e della Diffusione Intra Industriale e Territoriale dell'Innovazione Tecnologica", in Camagni R., Cappellin R. and Garofoli G. (eds.) **Cambiamento Tecnologico e Diffusione Territoriale**, Franco Angeli Editore, pp. 19-86

Monk P. (1989), **Technological Change in the Information Economy**, Pinter Publishers, London

Mueller J. (1991), "The European International Market for Telecommunications", **European Economic Review**, pp. 496-503

Myrdal G. (1959), **Teoria Economica e Paesi Sottosviluppati**, Feltrinelli, Milan

Nambu T. (1986), **Telecommunication Economics**, Nihon Keizai Schimbun, Tokio

Nash J. (1950a), "The Bargaining Problem", **Econometrica**, vol. 18, pp. 155-162

Nash J. (1950b), "Equilibrium Points in n-Persons Games", **Proceedings of the National Academy of Science**, vol. 36, pp. 48-49

Nash J. (1952), "Two-person Cooperative Games", **Econometrica**, vol. 21, pp. 128-140

Nelson R. and Winter S. (1977), "In Search of a Useful Theory of Innovation", **Research Policy**, vol. 6, pp. 36-76

Nelson R. and Winter S. (1982), **An Evolutionary Theory of Economic Changes**, Harvard University Press, Cambridge, Massachussets

Nijkamp P. (1977), **Theory and Application of Environmental Economics**, North-Holland, Amsterdam

Nijkamp P. (ed.) (1986a), **Handbook of Regional and Urban Economics**, North-Holland, Amsterdam

Nijkamp P. (ed.) (1986b), **Technological Change, Employment and Spatial Dynamics,** Springer-Verlag, Berlin

Nijkamp P. (1987), "New Technology and Regional Development", in Vasko T. (ed.), **The Long Wave Debate**, Springer Verlag, Berlin, pp. 274-282

Nijkamp P. and Priemus H. (1992), "Infrastructure and Network Access", mimeo Free University of Amsterdam

Nijkamp P. and Poot. J. (1993), "Endogenous Technological Change, Innovation Diffusion and Transitional Dynamics in a Non-Linear Growth Model", **The Australian Economic Papers**, forthcoming

Nijkamp P. and Reggiani A. (1992), **Interaction, Evolution and Chaos in Space**, Springer Verlag, Berlin

Nijkamp P. and Salomon I. (1989), "Future Spatial Impacts of Telecommunications", **Transportation Planning and Technology**, vol. 13, pp. 275-287

Nijkamp P. and Vleugel J. (1993), "Missing Networks and European Telecom Systems", in Bakis H., Abler R. and Roche M. (eds.), **Corporate Networks, International Telecommunicatios and Interdependence**, Belhaven Press, London, pp. 77-98

Nijkamp P., Leitner H. and Wrigley N. (eds.) (1985), **Measuring the Unmeasurable** Martinus Nijhoff, Dordrecht

Nijkamp P., Vleugel J., Maggi R. and Masser I. (1994), **Missing Transport Networks in Europe**, Avebury, Aldershot

Noam E. (1987), "The Public Telecommunications Network: a Concept in Transition", Columbia Institute for Tele-information, Columbia Business School, Working paper Series

Noam E. (1988), "Pluralism of Networks and Pluralism in Regulation: the Next Issues in Network Interconnection", Columbia Institute for Tele-Information, Columbia Business School, Working paper Series

Noam E. (1992), "Network Tipping and the Tragedy of the Common Network: a Theory for the Formation and Breakdown of Public Telecommunications Systems", in Antonelli C. (ed.), **The Economics of Information Networks**, North-Holland, Amsterdam, pp. 107-128

Norton D. and Rees J. (1979), "The Product Cycle and the Spatial Decentralisation of American Manufacturing", **Regional Studies**, n. 13, pp. 141-151

Oakey R. (1984), "Innovation and Regional Growth in Small High Technology Firms: Evidence from Britain and the U.S.A.", **Regional Studies**, vol. 18, pp. 237-251

Oakey R., Thwaites A. and Nash P. (1980), "The Regional Distribution of Innovative Manufacturing Establishments in Britain", **Regional Studies**, vol. 14, pp. 235-253

Oakland W. (1972), "Congestion, Public Goods and Welfare", **Journal of Public Economics**, vol. 1, pp. 339-357

OECD (1988a), **New Telecommunications Services: Videotex Development Strategies**, Paris

OECD (1988b), **The Telecommunications Industry: the Challenges of Structural Change**, Paris

Oniki H. (1990), "On the Cost of Deintegrating Information Networks", paper of the Institute of Social and Economic Research, mimeo

Ouwersloot H. (1994), **Information and Communications in an Economic Perspective**, Ph.D. Thesis, Free University, Amsterdam

Pasini G. (1959), **Impianti Telefonici: Criteri di Progettazione Razionale nella Telefonia Moderna**, Hoepli, Milan

Pavitt K. (1984), "Sectoral Patterns of Technical Change: Towards a Taxonomy and a Theory", **Research Policy**, vol. 13, pp. 343-373

Peschel K. (ed.) (1989), **Infrastructure and the Space Economy**, Springer Verlag, Berlin

Perez C. (1985), "Microelectronic, Long Waves and Structural Change: a World Case", **World Development**, vol. 13, n. 3, pp. 441-464

Perroux F. (1955) "Notes sur la Notion de Pole de Croissance", **Economie Appliquée**, n. 7, pp. 307-320

Philip G. (1990), "The Deregulation of Telecommunications in the EC", **International Journal of Information Management**, n. 1, march, pp. 67-75

Phillips A. (1990), "Changing Markets and Institutional Inertia: a Review of U.S. Telecommunications Policy", paper presented at the 8th International Conference of the International Telecommunications Society on "Telecommunications and the Challenge of Innovation and Global Competition", held in Venice, 18-21 March

Porat M. (1977), The Information Economy: Definition and Measurement, Special publications 77.22 (1), Office of Telecommunications, US Department od Commerce, Washington D.C.

Porter M. (1990), **Competitive Advantage of Nations**, Macmillan, London

Postrel S. (1990), "Competing Networks and Proprietary Standards: the Case of Quadraphonic Sound", **The Journal of Industrial Economics**, vol. 39, n. 2, pp. 169-185

Pred A. (1977), **City-Systems in Advanced Economies. Past Growth, Present Processes and Future Development Options**, Hutchinson, London

Pred A. and Tornqvist G (eds.) (1981), **Space and Time in Geography**, The Royal University of Lund, Lund

Quevit M. (1990), "Regional Technology Trajectories and European Research and Technology Development Policies", in Ciciotti E. Alderman N. and Thwaites A. (eds.) (1990), **Technological Change in a Spatial Context**, Springer Verlag, Berlin, pp. 317-338

Rabenau von B. and Stahl K. (1974), "Dynamic Aspects of Public Goods: a Further Analysis of the Telephone System", **The Bell Journal of Economics and Management Science**, vol. 5, n. 2, pp. 651-669

Reggiani A. (ed.) (1985), **Territorio e Trasporti: Modelli Matematici per l'Analisi e la Pianificazione**, Franco Angeli, Milano

Rietveld P. (1987), "Labour Supply for Additional Activities; a Micro Economic Approach", **International Journal of Development Planning Literature**, vol. 2, n. 1, pp. 29-53

Rietveld P. (1989), "Infrastructure and Regional Development: a Survey of Multiregional Economic Models", **The Annals of the Regional Science**, vol. 23, pp. 255-274

Rietveld P. (1990), "Employment Effects of Changes in Transport Infrastructure: Methodological Aspects of the Gravity Model", **Papers of the Regional Science Association**, Vol. 66, pp. 19-30

Rietveld P. and Nijkamp P. (1992), "Transport and Regional Development", **Series Research Memorandum**, Free University, Amsterdam

Rietveld P., Rossera F. and van Nierop J. (1993), "Technology Substitution and Diffusion in Telecommunication: the Case of the Telex and Telephone Network", **Sistemi Urbani**, forthcoming

Rodwin L. and Sazanami H. (eds.) (1991), **Industrial Change and Regional Economic Transformation: The Experience of Western Europe**, Harper Collins Academic, London

Rogers E. (1986), **Communication Technology: the New Media in Society**, the Free Press, New York

Rogers E. (1990), "The Critical Mass in the Diffusion of Interactive Technologies", paper presented at the 8th ITS Conference, held in Venice, 18-24 March

Rohlfs J. (1974), "A Theory of Interdependent Demand for a Communication Service", **The Bell Journal of Economics and Management Science**, n. 5, pp. 16-37

Roson R. (1993), "Spatial Implications of Network Externalities", mimeo available from the author, Dipartimento di Scienze Economiche, Università di Venezia Cà Foscari, Venice

Rullani E. and Zanfei A. (1988), "Networks between Manufacturing and Demand: Cases from Textile and Clothing Industry", in Antonelli C. (ed.) (1988), **New Information Technology and Industrial Change: the Italian Case**, Kluwer Academic Publisher Books, New York, pp. 57-96

Samuelson P. (1954), "The Pure Theory of Public Expenditure", **Review of Economics and Statistics**, vol. 36, pp. 387-389

Sandler T. and Tschirhart J. (1980), "The Economic Theory of Clubs: An Evaluative Survey", **Journal of Economic Literature**, vol. 18, December, pp. 1481-1521

Saraceno P. (1981), "Lo Sviluppo dell'Economia Meridionale all'Inizio del Nuovo Ciclo di Intervento Straordinario", **Informazioni Svimez**, n. 6, pp. 275-292

SAS / STAT (1989), **User's Guide**, Version 6, 4th Edition, Volume 1, Cary, SAS Institute Inc.

Saunder R, Warford J. and Wellenius B. (1983), **Telecommunications and Economic Development**, The Johns Hopkins University Press, London

Scherer F. (1980), **Industrial Market Structure and Economic Performance**, Rand McNally, Chicago

Schumacher E. (1965), **Social and Economic Problems Calling for the Development of Intermediate Technologies**, Abacus, London

Sciberras E. and Payne B. (1986), **Telecommunications Industry**, Longman, London

Scitovsky T. (1954), "Two Concepts of External Economies", **Journal of Political Economy,** vol. 62, pp. 143-151

Spence M. (1981), "The Learning Curve and Competition", **Bell Journal of Economics**, vol. 12, n. 1, Spring, pp. 49-70

Squire L. (1973), "Some Aspects of Optimal Pricing for Telecommunications", **Bell Journal of Economics and Management Science**, vol. 4, n. 2, Autumn, pp. 515-538

Stern E. and Krakover S. (1993), "The Formation of a Composite Urban Image", **Geographical Analysis**, vol. 25, n. 2, pp. 130-146

Stöhr W. (1986), "Territorial Innovation Complexes", in Aydalot P. (ed.), **Milieux Innovateurs en Europe**, GREMI, Paris, pp. 29-56

Stöhr W. and Tödtling F. (1977), "Spatial Equity - Some Anti- Theses to Current Regional Development Doctrine", **Regional Science Association Papers**, vol. 38, pp. 33-53

Stoneman P. (1983), **The Economic Analysis of Technological Change**, Oxford University Press, Oxford

Stoneman P. (1986), "Technological Diffusion: The View Point of Economic Theory", **Ricerche Economiche**, vol. 4, pp. 585-606

Stoneman P. (1990), "The Intertemporal Demand for Consumer Technologies requiring Joint Hardware and Software Inputs", working paper n. 355, University of Warwick, Coventry CV4 7AL, Great Britain

Stoneman P. (1992), "The Impact of Technology Adoption on Firm Performance: Heterogenity and Multi-Technology Diffusion Models", mimeo

Swyngedouw E. (1987), "Social Innovation, Production Organisation and Spatial Development", **Revue d'Economie Régionale et Urbaine**, n. 3, pp. 487-510

Thwaites A. and Alderman N. (1990), "The Location of Industrial R&D: Retrospect and Prospect", in Cappellin R. and Nijkamp P. (eds.), **The Spatial Context of Technological Development**, Avebury, Aldershot, pp. 17-42

Tirole J. (1988), **The Theory of Industrial Organisation**, The MIT Press, London and Cambridge, Massachusetts

Tolmie I. (1987), "The Technological and Organisational Base of Computer Networks", mimeo available from the author, Centre for Urban and Regional Development Studies, University of Newcastle, Newcastle upon Tyne, NE1 7RU

Townroe P. (1990), "Regional Development Potentials and Innovation Capacities", in Ewers H. and Allesch J. (eds.) **Innovation and Regional Development**, de Gruyter, Berlin, pp. 71-82

Utterbach J. (1974), "Innovation in Industry and Diffusion of Technology", **Science**, pp. 620-626

Varian H. (1992), **Microeconomic Analysis**, Norton, 3th Edition, New York

Vernon R. (1957), "Production and Distribution in the Large Metropolis", **Annales of the American Academy of Political and Social Sciences**, pp. 15-29

Vernon R. (1966), "International Investment and International Trade in the Product Cycle", **Quarterly Journal of Economics**, May, vol. 80, pp. 190-207

Vickerman R. (ed.) (1991), **Infrastructure and Regional Development**, Pion Limited, London

Von Weizsaecker C.C. and Wieland B. (1988), "Current Telecommunications Policy in West Germany", **Oxford Review of Economic Policy**, vol. 4, n. 2, pp. 20-39

Weisberg S. (1980), **Applied Linear Regression**, John Wiley and Sons, New York

Wenders J. (1987), **The Economics of Telecommunications**, Ballinger

Williams H. (1987), "The Use and Consequences of Information and Communications Technologies and Trade in Information Services", mimeo available from the author, Centre for Urban and Regional Development Studies, University of Newcastle, Newcastle upon Tyne, NE1 7RU

Williams H. and Gillespie A. (1989), "A Small Firm Perspective on the Liberalisation of Telecommunications Services", in Garnham N. (ed.) **European Telecommunications Policy Research**, Springfield Publisher, Amsterdam, pp. 183-202

Williams H. and Taylor J. (1991), "ICTs and the Management of Territory", in Brotchie J., Batty M., Hall P. and Newton P. (eds.), **Cities of the 21st Century**, Halstead Press, Longman, Cheshire, pp. 293-306

Williamson O. (1975), **Markets and Hierarchies: Analysis and Antitrust Implications**, The Free Press, New York

Willinger C. and Zuscovitch E. (1988), "Towards the Economics of iInformation-Intensive Production Systems: the Case of Advanced Materials", in Dosi G., Freeman C., Nelson R., Silverberg G. and Soete L. (eds.), **Technical Change and Economic Theory**, Pinter Publisher, London, pp. 239-255

Zanfei A. (1990), **Complessità e Crescita Esterna nell'Industria delle Telecomunicazioni**, Franco Angeli, Milan

Zeleny M. (1985), "La Gestione a Tecnologia Superiore e la Gestione della Tecnologia Superiore", in Bocchi G. and Ceruti M. (eds.), **La Sfida della Complessità**, Feltrinelli, Milan